MATHEMATICAL INTUITIONISM AND INTERSUBJECTIVITY

SYNTHESE LIBRARY

STUDIES IN EPISTEMOLOGY,

LOGIC, METHODOLOGY, AND PHILOSOPHY OF SCIENCE

Managing Editor:

JAAKKO HINTIKKA, *Boston University*

Editors:

DIRK VAN DALEN, *University of Utrecht, The Netherlands*
DONALD DAVIDSON, *University of California, Berkeley*
THEO A.F. KUIPERS, *University of Groningen, The Netherlands*
PATRICK SUPPES, *Stanford University, California*
JAN WOLEŃSKI, *Jagiellonian University, Kraków, Poland*

VOLUME 279

TOMASZ PLACEK
*Jagiellonian University,
Cracow, Poland*

MATHEMATICAL INTUITIONISM AND INTERSUBJECTIVITY

A Critical Exposition of Arguments for Intuitionism

KLUWER ACADEMIC PUBLISHERS
DORDRECHT / BOSTON / LONDON

A C.I.P. Catalogue record for this book is available from the Library of Congress.

ISBN 0-7923-5630-6

Published by Kluwer Academic Publishers,
P.O. Box 17, 3300 AA Dordrecht, The Netherlands.

Sold and distributed in North, Central and South America
by Kluwer Academic Publishers,
101 Philip Drive, Norwell, MA 02061, U.S.A.

In all other countries, sold and distributed
by Kluwer Academic Publishers,
P.O. Box 322, 3300 AH Dordrecht, The Netherlands.

Printed on acid-free paper

All Rights Reserved
© 1999 Kluwer Academic Publishers
No part of the material protected by this copyright notice may be reproduced or
utilized in any form or by any means, electronic or mechanical,
including photocopying, recording or by any information storage and
retrieval system, without written permission from the copyright owner.

Printed in the Netherlands.

What a mathematician is inclined to say about the objectivity and reality of mathematical facts, is not a philosophy of mathematics, but something for philosophical treatment.

        Wittgenstein, *Philosophical Investigations*, §254

TABLE OF CONTENTS

ACKNOWLEDGMENTS   xi

WORKS FREQUENTLY QUOTED   xii

CHAPTER 1   Introduction   1
  1   Objectives of this Study   1
  2   Intersubjectivity and Conditions of Intersubjectivity   4
  3   Mathematicians on Intersubjectivity   12

CHAPTER 2   Brouwer's Philosophy   17
  1   The Knowing Subject   18
    1.1   Mind, Time and Objects   18
    1.2   Brouwer and the Problem of Other Minds 22
    1.3   The Basic Intuition of Two-ity   27
  2   Mathematics and Intuition   29
    2.1   Infinitely Proceeding Sequences, Spreads and Species   29
    2.2   What Is Not Intuitionistically Intuitive?   36
    2.3   Some Objections to Brouwer's Concept of Intuition   40
    2.4   The *Possible* Construction: Brouwer's Notion of Possibility   44
  3   Language, Truth and Relations Between Logic and Mathematics   48
    3.1   Language   48
    3.2   Against Hilbert's Program   51
    3.3   Brouwer's and the Axiomatic-Deductive Method   54
    3.4   What Is Logic?   59
    3.5   Mathematics vs. Logic   62
    3.6   Against *Begriffe*   65
    3.7   What Is Truth?   67

	3.8	The Validity of Laws of Logic	69
		3.8.1 Weak Counterexamples	69
		3.8.2 A Reconstruction of Brouwer's Argument	71
		3.8.3 Indeterminacy or Infinity	74
		3.8.4 Strong Counterexamples to the (Generalized) Excluded Middle	79
4	Intersubjectivity in Brouwer's Conception of Mathematics		83
	4.1	Psychologism, Subjectivism, Solipsism?	84
	4.2	Brouwer and Intersubjectivity: the Mentalist Condition	89
	4.3	How Can One Communicate about Mental Constructions?	90
5	Conclusions about Brouwer's Philosophy		100

CHAPTER 3 Heyting's Arguments — 103

1	Against Intuitionistic Philosophy, for Intuitionistic Psychology?	104
2	Intuition as Self-Evidence	108
3	The Neutrality Argument	112
	3.1 Heyting's Counterexamples: What Do They Prove?	113
	3.2 From Ontological Neutrality to the Repudiation of Bivalent Truth	119
4	The Semantical Argument	126
	4.1 A Note on Heyting's Views on Formalization and Logic	137
5	Intersubjectivity in Heyting's Conception	139
6	Resume of Heyting's Arguments	144

CHAPTER 4 Dummett's Case for Intuitionism — 147

1	Dummett's Program: An Overview	147
2	Dummett on Semantic Theories	151
	2.1 Three Tasks of Semantic Theories	151
	2.2 Programmatic Interpretation	155
	2.3 Skeletal Semantics	161
3	Dummett on Meaning and its Basis	164
	3.1 Meaning, Knowledge and Understanding	164
	3.2 Sense, Force and Holism	170
	3.3 Sense and Semantic Theory	173
4	The Language-Learning Argument	176
	4.1 Knowledge of Truth Conditions	176

4.2	The Ingredient of Meaning That Transcends Use	181
4.3	Why Intuitionistic Provability?—Holism to the Rescue	187
5	Resume of Dummett's Argument	192

CHAPTER 5 Conclusions   194

APPENDIX   197

NOTES   203

BIBLIOGRAPHY   207

INDEX   213

ACKNOWLEDGMENTS

The present book is a thoroughly revised and extended version of the doctoral dissertation that I wrote under the supervision of Professor Jan Woleński and defended at the Jagiellonian University in Cracow in 1991. I am indebted to Jan Woleński for both his valuable comments on various drafts of the work and his permanent encouragement first for the project and then for its preparation for publication. I also wish to acknowledge my indebtedness to Professor Józef Misiek who sparked off my interest in the philosophy of mathematics and intuitionism. The book owes much to my discussions with friends and colleagues as well as to their comments on and stimulus for my work. Thus, Paweł Turnau's provocative questions concerning Brouwer's and Heyting's conceptions of mathematics have made me thoroughly rethink and rewrite the respective chapters. Anna Kanik read the material about Brouwer's and Heyting's philosophy of mathematics and suggested a number of improvements. Artur Rojszczak helped me to clarify mysteries of Husserl's theory of meaning and Eugeniusz Tomaszewski commented on my presentation of a strong counterexample to the excluded middle. Katarzyna Kijania-Placek thoroughly and accurately read the whole manuscript and suggested many corrections, especially in the presentation of mathematical reasonings. Finally, William Brand not only corrected the English of this work, but also suggested clarification of the philosophy it contains. Some material from the book has also been included in talks I gave at the 13th Wittgenstein Symposium in Kirchberg and at the Boston Colloquium of Philosophy of Science. I am indebted to the disputants. Finally, I acknowledge the financial help of the Jagiellonian University in Cracow, Poland, a grant from which enabled me to have the language of the manuscript corrected and the book typeset.

T.P.

WORKS FREQUENTLY QUOTED

The list below includes titles of works that are often referred to in the book together with the abbreviations used. With these exceptions, works are referred to in the usual way, that is by giving the name of the author, year of publication and (if needed) the page number. If no confusion is likely, however, the author's name is dropped.

TE	M. Dummett: *Truth and Other Enigmas*
LBM	M. Dummett: *The Logical Basis of Metaphysics*
PBIL	M. Dummett: 'The Philosophical Basis of Intuitionist Logic'
WTM	M. Dummett: 'What Is a Theory of Meaning II'
EI	M. Dummett: *Elements of Intuitionism*
LU	E. Husserl: *Logische Untersuchungen*
VS	W.P. van Stigt: *Brouwer's Intuitionism*

1
INTRODUCTION

In 1907 Luitzen Egbertus Jan Brouwer defended his doctoral dissertation on the foundations of mathematics and with this event the modern version of mathematical intuitionism came into being. Brouwer attacked the main currents of the philosophy of mathematics: the formalists and the Platonists. In turn, both these schools began viewing intuitionism as the most harmful party among all known philosophies of mathematics. That was the origin of the now-90-year-old debate over intuitionism. As both sides have appealed in their arguments to philosophical propositions, the discussions have attracted the attention of philosophers as well.

One might ask here what role a philosopher can play in controversies over mathematical intuitionism. Can he reasonably enter into disputes among mathematicians? I believe that these disputes call for intervention by a philosopher. The three best-known arguments for intuitionism, those of Brouwer, Heyting and Dummett, are based on ontological and epistemological claims, or appeal to theses that properly belong to a theory of meaning. Those lines of argument should be investigated in order to find what their assumptions are, whether intuitionistic consequences really follow from those assumptions, and finally, whether the premises are sound and not absurd. The intention of this book is thus to consider seriously the arguments of mathematicians, even if philosophy was not their main field of interest. There is little sense in disputing whether what mathematicians said about the objectivity and reality of mathematical facts belongs to philosophy, or not. I agree with Wittgenstein, however, that it is a task of philosophy to study and investigate their arguments.

1. OBJECTIVES OF THIS STUDY

Why, however, are we concerned with the problem of the intersubjectivity of mathematical knowledge? The immediate reason for investigating this issue is the mutual allegations by proponents of intuitionism and classical mathematics that their opponent's philosophy of mathematics leads to the conclusion that mathematical knowledge is not communicable.

On the one hand, it is often remarked about Brouwer's and Heyting's justifications of intuitionism that if mathematics were practiced in the way they prescribed, mathematicians could not communicate their results, check other persons' proofs, or successfully teach their subject.[1] In a somehow similar vein, their conception of mathematics is accused of giving rise to 'subjectivism', 'psychologism' or even, 'solipsism'. Also, it is sometimes alleged that it is hard to say what in this conception of mathematics allows for (or guarantees) the intersubjectivity of knowledge. These objections can be traced back to two basic theses held by the 'traditional intuitionists', Brouwer and Heyting:

T1:
Mathematical objects are mental constructions. (Heyting 1956, p. 1) These constructions are completely independent from language as they originate in 'the self-unfolding intuition of two-ity'. (Brouwer 1933, p. 443) Properties of mathematical objects are not determined by logic, as mathematics is prior to it.

The above thesis suggests that the status of mathematical constructions is somehow 'private'. The second thesis adds that:

T2:
The exactness of mathematics cannot be secured by linguistic means. (Brouwer 1933, p. 443) This is because language is an irreparably imperfect means of communication and inevitably inadequate as a mode of description of mathematical constructions. (Brouwer 1952b, p. 510) Language is basically a vehicle of the transmission of will. In this transmission there is neither exactness, nor certainty. The same imperfection is inherent in the language of mathematical logic, for logic deals with linguistic phenomena only; the rules of logic merely describe regularities obtaining between assertions. Thus, the formalist school is mistaken in its attempts to secure the certainty of mathematics by the formalization of its language. (Brouwer 1929, pp. 421–2)

On the other hand, a somewhat similar objection to classical semantics has recently been raised by Michael Dummett. He holds that a theory of meaning based on the concept of bivalent truth leads to the conclusion that knowledge is not communicable. As the logical operators in classical mathematics are defined in terms of truth, the ultimate target of Dummett's attack is classical mathematics.

Observing such mutual accusations one obviously wonders whether they are justified. Clearly, discovering that the objection in one or the other case is correct will undermine the position of the school objected to, at least to the degree that a given justification for that conception of mathematics should then be discarded and replaced by a better one. In an attempt to find an answer to this problem, we shall bring under scrutiny the arguments for replacing classical mathematics by intuitionistic mathematics.

The second problem studied here arises from noting that the intuitionists' attacks on classical mathematics often give the impression that their central claims concern the *meanings* of mathematical theorems. Both Brouwer and Heyting sometimes write as if they were arguing that statements of classical mathematics had no meaning, or that they received meaning only if they were interpreted in the intuitionistic manner. These formulations suggest a similarity between their reasoning and that of Dummett, who argues, taking as a starting point some considerations belonging to theories of meaning and semantics, that logical connectives and quantifiers cannot have the meanings that classical mathematicians or logicians ascribe to them. Now, according to some prominent thinkers like Frege, Tarski or Davidson, the concepts of 'truth' and 'meaning' are closely related, as one is definable in terms of the other. Thus, to define truth in a Tarskian style one needs to appeal to the notion of translation, a close relative to that of meaning. On the other hand, given that we know what truth is, we may explicate the concept of meaning, as Davidson does. These sorts of considerations may incline one to think that the intuitionists had an account of meaning applicable perhaps not to the entire language, but at least to the language of mathematical discourse, and that this account forced them to reject bivalent truth, resulting in turn in the rejection of some theorems and means of proof commonly accepted in mathematics. Thus, we shall ask what concept of meaning the intuitionists had (if any), whether their ideas on meaning are sound, and whether they serve as a rationale for accepting basic intuitionistic tenets, like intuitionistic constructibility or the rejection of bivalence.

Finally, the last objective of this book: we shall investigate justifications of mathematical intuitionism, that is, rationales for a revision of classical mathematics and its replacement by intuitionistic mathematics. For that very reason, various 'eclectic' philosophies of mathematics are not taken into account. As the term 'eclectic' suggests, such philosophies assume that classical as well as intuitionistic mathematics are equally correct and equally worth practicing. Leaving aside such 'eclectic' justifications of intuitionism, we will be concerned with evaluating the arguments for militant intuitionism, that is for a position claiming that classical mathematics is essentially flawed and needs to be revised in the intuitionistic manner. We will be interested to know if these arguments are formally correct and if their premises are tenable. That will shed some light on whether mathematical intuitionism is a philosophically viable position. In what follows, three arguments in favour of mathematical intuitionism are examined, that is, Brouwer's, Heyting's and Dummett's. We concentrate on these reasonings simply because they are best known at present and although other theories, most notably Per Martin-Löf's meaning-theory for constructive mathematics, highly deserve examination, we postpone this research to some future work, fearing that including more material would endanger the integrity of the present book.[2]

To sum up, we shall study three justifications of intuitionism here in order

1. to find out whether the mutual accusations of the incommunicability of mathematical knowledge are justified;
2. to investigate whether the intuitionists had an account of meaning, and if so, whether there is any relation between the assumed concept of meaning and the concept of truth they accept;
3. to assess how convincing the intuitionists' rationales for the revision of mathematics are.

2. INTERSUBJECTIVITY AND CONDITIONS OF INTERSUBJECTIVITY

We will be discussing below those arguments for and against the intuitionistic program that invoke the issue of intersubjectivity. Some touch only indirectly on this problem in that they give rise to doubts about how mathematics could be communicated if a given picture of mathematics were correct. Others are more directly linked to this issue. Thus, to get a grip on whether or not these arguments are correct, we need to become clear about the notion of intersubjectivity.

The notion of intersubjectivity is closely related to that of the possibility of communication. It is perhaps too strong a requirement to demand that one should be able to communicate everything that one can be aware of. Therefore, a distinction is drawn between thoughts, which are assumed to be mental objects that bear knowledge, and other sorts of mental objects. The next step is to claim that intersubjectivity concerns only thoughts: they alone should be conveyable or communicable. For instance, we do not require that a hardly verbalizable fear or a sensation be communicable, although someone's thought that he has a sensation should be conveyable. Since in this view a thought always belongs to someone's psychic inventory, the idea that thoughts should be conveyed cannot be understood literally, as a transmission of numerically the same object. Rather, people need to have the ability, when appropriately prompted, to form similar thoughts. In this view, the question of what constitutes the assumed similarity of thoughts as well as the basis for discerning thoughts from among mental objects is usually left unanswered. Nevertheless, this picture easily yields the following condition for intersubjectivity:

THE MENTALIST CONDITION
For two people to communicate, their perceptual contents, thoughts or memories should contain some invariant elements, common to all thinking human

subjects. There should be such invariants in spite of individual differences among people and the changes they go through in their lives. The condition requires that the theory of knowledge permit such invariant elements. (Klaus and Buhr 1972, p. 258)

According to the above condition, I manage to communicate with another person about a given subject matter because our thoughts (perceptual contents, memories), despite being different, contain the same ingredients. In a similar vein, I understand what I once wrote down since the thought that I once believed contained the same invariant elements as the thought that originates in my mind while I am reading my old note.

It is plain that the above picture of human communication ignores the means by which we communicate. Since we are preoccupied with the intersubjectivity of knowledge, and knowledge can hardly be communicated in any other medium than language, we may limit our attention to communication through language. Accordingly, the task of explaining intersubjectivity amounts to finding out what constraints a model of language must assume in order to make linguistic communication possible. Now, given the standard distinction between what a speaker says by uttering a given expression and the point of his utterance of the expression, the question arises whether intersubjectivity should concern both features or only the former. The decision, as we see it, is to some extent arbitrary and depends on what one needs the concepts of intersubjectivity or communication for. Our objective is to have a viable concept of the communication of mathematical, or more generally, scientific knowledge. Thus, our paradigmatic examples involve a language-user who hears someone uttering a sentence, or reads a sentence and understands what the sentence *expresses*. Accordingly, we assume that if a person misconceived the speaker's intention while grasping what was said, this does not mean that the uttered sentence was not intersubjective. To take an example, the remark 'Napoleon died on St. Helena Island' uttered sardonically during a party conversation that dwelt on the host's cat, named Talleyrand, who is in an animal hospital incidentally known as 'St. Helena' is *understood* by a person who takes it as expressing a piece of historical information, but is totally baffled about why the remark fell at that point in the conversation and what the speaker's intention was. However, if one has a different end in view, let us say that of accounting for people's understanding of each other's intentions, then he should perhaps take other features into account as well, beyond what the speaker says in uttering a given expression.

How one further elaborates on the equating of intersubjectivity with understanding uttered expressions in the same way, crucially depends on what categories are assumed to be central to the account of language one prefers. A wide class of such accounts chooses denotation as just such a central category. Now, if we first concentrate on extensional contexts of language and leave the non-extensional ones for special treatment, and in the next step identify denotations of individual terms with their bearers and denotation of sentences with their truth-values, we will be following in Frege's footsteps. To develop further this account along Fregean lines we would obviously have to introduce the concept of the *sense* of an expression as that part of understanding the expression that provides us with a means, not necessarily effective, of determining the denotation of the expression. Still further, we would need to explain, partly in terms of both denotations and senses, various basic uses of sentences, such as making assertions, issuing commands, or asking questions. This will fall under the heading of a theory of force. But we naturally hope, when bringing our problem of intersubjectivity into confrontation with such a pyramid-like structure as a Fregean model of language, that the problem can be addressed at the very first level. Hence we ask whether it is not enough for successful communication that the individuals involved use expressions that have the same denotations, regardless of any differences among the senses that they ascribe to these expressions. A paradigmatic example of such a situation will be a conversation about a person known to the parties involved as 'Smith', but identifiable by each of them by completely different means and methods. To say that their methods of identification are completely different is most easily represented by assuming that the set of beliefs of the form 'Smith is ψ' that the first speaker has does not overlap with the respective set of the second speaker. This does not mean that the beliefs of the first speaker contradict those of the other; even more, given that sense determines denotation, they cannot contradict each other because otherwise if only all these beliefs were true, the name 'Smith' would denote different objects for the first and the second speaker. We should rather view these sets in such a way that the information contained in only the first set does not allow one to derive any of the beliefs of the second set and the other way round. Now, if each speaker assumes different statements to be true of Smith, can they communicate anything about him? To take an example, will the utterance by the first speaker of the sentence 'Smith has bought a new bike' be understood by the second speaker as conveying information concerning Smith? If there is a temptation to say 'yes', it stems from ignoring the possibility that the speakers may form a suspicion that despite using the same name they are talking about different persons or objects. If only such a suspicion arises, however, given there is no overlap among the speakers' set of beliefs, the speakers will not be able easily to resolve it, since in order to establish that they are talking about the same person they need to acquire an additional piece of information, over and above the beliefs they hold. As long as they do not have this sort of information, they cannot be said to have communicated because they cannot resolve their possible perplexity as to who the conveyed information is about. Moreover, we have so far assumed that both the speakers differ only in as much as they ascribe different senses to the word functioning as the subject of the uttered sentence. If, however, identity of denotations is all that one demands for understanding, then we should assume as well that the speakers understand all other expressions differently. It is clear that as soon as they entertain the slightest doubt as to whether they are really speaking of the same matter, they will land in an almost insoluble perplexity. Thus, what emerges from this discussion is a requirement of some constraints on how much the sets of beliefs concerning a given object and accepted by both the speakers may diverge for the speakers to be able to find out whether they speak of the same thing, whenever this doubt arises. We may express this in Frege's terms by saying that there should be some constraints on the extent to which senses ascribed to a given expression may diverge. Denotations can serve to formulate only a negative criterion: given that two speakers ascribe different denotations to the same expression, they cannot communicate by using this expression.

To what extent then should the senses that people ascribe to the expression overlap in order for the communication be possible? Clearly, to demand a complete overlap is far too much. If we think of senses ascribed as represented by sets of beliefs concerning the object involved, should we say that there must be some common beliefs that moreover suffice for uniquely determining the denotation of an expression in question? Again, to require such a common core of beliefs that are actually held by the speakers seems to be unrealistic. Rather, we say that a message is intersubjective if one person, acting exclusively on his beliefs, may derive some beliefs from the set of the other speaker so that they suffice for uniquely determining the denotations of the expressions involved. The word 'exclusively' is meant to exclude the possibility of consulting an external source of information, like looking into an encyclopedia, asking an expert on the subject, or finding the intended denotation and showing it to the other speaker. Why do we demand, however, that the derivation be *exclusively* based on the set of accepted beliefs? A motivation to do so comes from realizing that once we agree that in finding out the denotation of an expression one may draw on any possible information whatsoever, then this explanation becomes virtually empty. Thus, in order to guard our account against this degeneration, we need to put some restriction on possible information. To consider linguistic communication, at the one extreme we have two speakers that associate with their common expression α sets ψ_1 and ψ_2 of their beliefs, respectively and such that some beliefs from ψ_2 that determine the denotation of α can be derived solely from the beliefs from the set ψ_1. In the middle we have such differences of the senses ascribed as make the derivation impossible; nevertheless, there is an effective method of finding what the expression stands for. We may for instance think of two speakers conversing about Rome, of whom the first knows only that Rome is the capital of Italy, while the other is ready to assert only that the Pope resides in Rome. Now, given encyclopedias, geographical dictionaries, tourist guides and the like, they can easily find out that they are talking about the same city. Finally, at the other extreme we have cases in which there is no such effective method, nor any guarantee that a selected method will prove successful. To give an example, suppose there is the name α of a historic figure of whom some historians have only the set ψ_1 of beliefs, while other historians accept only the beliefs of ψ_2, and moreover, no belief of ψ_1 can be derived solely from ψ_2 and vice versa. Still more, at the time of an utterance that attempts to convey something concerning α, no information is available such that, if added to set ψ_1, it would yield such beliefs of ψ_2 that uniquely determine the denotation of α, or the other way round. No encyclopedia, no historical sources, no memoirs decide whether the person of whom beliefs of ψ_1 are true is the same as the person identified by beliefs of ψ_2. In such circumstances, as the doubt whether the historians are speaking of the same person cannot be resolved, they do not communicate anything about this person. In a somewhat similar vein, we would rather hesitate to say that somebody's utterance is intersubjective if, despite there being a method to decide whether the sentence is about the same object, acquaintance with it and ability to use it require some ingenuity on the part of the hearer. It may be said in such cases that knowledge of this method far exceeds everything that may be reasonably demanded of the abilities of the average language-user. To illustrate, suppose that two persons chatting about one Mary X. realize that since their respective beliefs about Mary X. largely diverge, they might in fact be talking about different people. They surely have means, ranging from finding common acquaintances, consulting phone books or, in the last recourse, employing a detective agency, to find out whether they are talking about the same person. Nevertheless, to demand that they engage in such investigations is in most circumstances far too much. We would rather say that in such a case the communication is deemed a failure.

The discussion so far has shown that the stipulation of deriving appropriate beliefs of ψ_1 *exclusively* from those of ψ_2 and *vice versa* is too strong. We should rather reformulate our initial stipulation in the following way: people can communicate if one person, acting on the set of beliefs he accepts and collateral information that he may gain by following some routine practice, can derive such beliefs of the set of beliefs of the other speaker that suffice for uniquely determining the denotations of the expressions used in the uttered sentence. By 'routine practice' we mean here things like checking words in encyclopedias, dictionaries, handbooks, consulting experts acknowledged in a given area, pointing to some objects that are in the vicinity of the speakers, and the like.

It is hard to resist the impression that this characterization of intersubjectivity is far from workable. It assumes that a sense ascribed by someone to an expression can be represented as a set of his beliefs concerning a given object. It poses, moreover, the demanding task of deciding whether a given means of gaining collateral information about the denotation of an expression counts as routine. Instead of someone's beliefs, however, we may talk of that person's readiness to assert the corresponding sentences and suspend the assertion of some other sentences. The asserting of sentences is one of the principal aspects of the use of language. Thus, a temptation arises to formulate a condition of intersubjectivity entirely in terms of language-use, with a hope that the difficulty of delineating of routine practices will be circumvented. The turn from senses to use has yet another advantage. Until now we have been working in a tacitly assumed Fregean framework that involved dealing with three levels of explanation: denotations, senses, and forces of expressions. The decision to frame a condition of intersubjectivity in terms of assertion or other aspects of language-use leaves us neutral as to whether or not expressions *refer*. This is of some importance since, as it shall turn out, the idea that mathematical expressions refer brings troublesome consequences for Brouwer's and Heyting's conceptions of mathematics.

How then should the desired condition be formulated? We content ourselves with a negative statement to the effect that, if a model of language does not allow people to acquire essentially the same use of an expression, then communication is impossible. As this condition goes against a mentalist picture of analogous thoughts, mental contents and the like, and as it is Wittgenstein who is widely known for strongly inveighing against this view, let us call this requirement: *Wittgensteinian.*

WITTGENSTEINIAN CONDITION:
In the process of language-learning, people should be able to acquire abilities to use linguistic expressions in basically the same way. They should be able to learn in what circumstances one is entitled to assert a sentence, in what circumstances one may disagree with assertions of other speakers, what question should be asked for receiving the desired information, and many other equally important practices. The ability to use an expression in basically the same way is all that can be required of speakers who are supposed to communicate.

To be clear about it, the condition does not require speakers to use expressions in exactly the same way. Two persons who diverge in their uses of the words 'mug' and 'cup' communicate in most situations. To allow for such idiosyncrasies, we say that expressions should be used in *basically* the same way.

Now, to elucidate the condition, let us inquire what views on language conflict with this requirement. We will deal with only two of them. The first was invoked by Dummett in his argument against truth-conditional meaning-theories. On the view Dummett criticizes, it is agreed that some linguistic practices, say of using a language of mathematical discourse, can be displayed to a novice, so that he will learn which computations establish numerical formulae, on what grounds a mathematical statement can be asserted, what are the immediate consequences of a given statement, and many other features of use. Nevertheless, the proponent of this view contends that *mere* mastery of language use is not enough for a person to understand an expression—he needs to acquire a grasp of some other feature that, moreover, cannot be manifested by displaying various aspects of the use of language. The consequence is that two mathematicians who competently discuss a mathematical proof but nevertheless lack that mysterious element of understanding have only an illusion of mutual understanding. Clearly, on this view the ability to use expressions in basically the same way does not suffice for two people to communicate.

The other position that may (though not necessarily) be at variance with our requirement assumes that expressions *refer* to mental objects. It may be suspected that this view on denotation is implied by Brouwer's and Heyting's doctrine of mathematical objects being mental constructions. We have there two ideas, that expressions refer and that their denotations are mental objects, the combination of which invites the question of how such objects are to be named, and this, as better candidates are lacking, lands us with the concept of private ostensive definition. Now, even if one comes up with a satisfactory answer about how the private naming is effected or how it happens that one *correctly* recognizes objects that he had earlier named privately, there is a remaining question of whether two speakers can understand the expression in the same way. On the one hand, if understanding involves people having the same 'mental denotations' of the expression involved, even if it is granted that different people's mental objects may be qualitatively the same, then speakers will understand the expression in an arbitrary way, since others have no control over what object the person names by a given expression. The view thus clearly violates the Wittgensteinian condition. On the other hand, to secure the satisfaction of the condition, one may drop the requirement of the same 'mental denotations', but this will result in questioning what all this talk of denotations, mental objects and private naming is for. We shall further elaborate on these problems when we discuss whether the Wittgensteinian condition is satisfied in Brouwer's and Heyting's conceptions of intuitionism.

The mentalist and Wittgensteinian requirements for intersubjectivity that we singled out above are in conflict, or a least a strong case may be put that they are of a conflicting nature. Accordingly, one may construct an account that satisfies one, but violates the other. The prevailing view, at least in the mainstream of the analytic philosophy, is that the mentalist condition is incorrect, as it stems from a rather naive view on human communication. Nevertheless, the second condition, or at least its second ingredient that equates agreement in use of the expression with agreement on its meaning, is also less than absolutely water-tight. Taking an analogy between language and the game of chess, one may cogently argue after Dummett (1991, p. 88) that mere observance of the rules and mastery of playing is still too little to credit a player with understanding the game. Rather, we would also demand that the player know the point of his moves, or know the fun of devising a strategy and observing that it works. With a little fantasy we may imagine with Dummett a player who automatically but brilliantly makes moves in chess, but neither devises a strategy nor is pleased nor disappointed by the outcomes of his moves. Most of us will probably be inclined to say that such an automaton does not *understand* chess. Our problem, however, is not as much the understanding of chess or a sector of language, as linguistic communication or *mutual* understanding. Accordingly, we should rather ask whether this automaton communicates with its chess-partner. Now, if only it is meaningful to talk about communication in chess, then clearly its main ingredient is the ability to counter the partner's moves with one's own correct moves, and that is precisely what we have assumed our player can do. Clearly, what the automaton cannot achieve, given our assumptions, is (1) to consciously devise a chess strategy, (2) to form a conception of why his partner has chosen just that move from among a number of possible ones, (3) to feel what a human player feels while, say, checkmating. Clauses (1) and (2) are rather akin, as they both hint at an ability to act for a reason and to ascribe reasons for another's action, whereas clause (3) suggests similarity with the mentalist condition, though as we shall see, they are not congruous. The clauses represents features that may go into understanding; accordingly, we need to decide which should be conveyed in communication.

Let us reflect first on what it is to say that our automaton cannot consciously devise a strategy in the game. This means that it is unable to reflect on a sequence of moves it makes while playing chess or, analogously, on a sequence of utterances on a given occasion. Now, given a strong suspicion that our partner in a discussion cannot reflect on his linguistic practice, even though his behavior during the conversation is otherwise perfectly average, are we inclined to say that we do not communicate? If we think of this possible doubt as arising solely from observation of linguistic behavior, then the question is on the verge of being meaningless, as it is hard to combine someone's perfectly unremarkable linguistic behavior during a conversation with such a suspicion. Things will look different, however, if we conceive of that doubt as drawing on other sorts of information, for instance concerning the physiological setup or psychology of the creature involved. To say it in the other words, if we meet a Martian that speaks our language, we may suspect that the communication is nevertheless illusory because either we observe that he is made of a different sort of stuff and start doubting whether he can *really* understand, say, our remarks about feats of eating, or learn that Martians do not have certain human feelings and accordingly, doubt if they can understand, say, our love poems. As easily noticed, it is exactly these doubts (i.e., concerning the creature's physiology or psychology) that lie behind the appeal of the mentalist condition. There is nevertheless a certain disparity between the proponent of the mentalist condition and someone who is skeptical about achieving communication with Martians on the grounds of their different psychology, as the former compares the thoughts of individuals while the latter is concerned with possible differences between the psychology of one group and that of another.

Stories about a chess-playing automaton, a Martian, or other unearthly creatures suggest a distinction between genuine communication and illusory communication with individuals who can perfectly obey the rules of our language without completely understanding the information. If this is so, we should relax the second ingredient of the Wittgensteinian condition and ascribe the same meaning of an expression to creatures who use the expression in basically the same way, as long as they belong to the same species. What the condition excludes is a concept of a factor that, no matter how well two persons (or two Martians) agree in the way they use relevant expressions, they cannot be said genuinely to communicate unless they have (presumably, in their mental inventory) the factor mentioned.

But do we really communicate? Clearly, this is the voice of a skeptic of a sort already known to the ancient Sophists. Upon encountering this character, it is time to realize how limited the objectives of the present work are. We had better point out to him a shelf with a long row of books on mathematics, including some of the intuitionistic variety, and say: *they* have communicated, even if the rest cannot. Thus, while believing that people communicate, we shall check whether the conditions on intersubjectivity that we singled out are satisfied in the philosophies of mathematics of Brouwer, Heyting, and Dummett.

3. MATHEMATICIANS ON INTERSUBJECTIVITY

Is the issue of the intersubjectivity of mathematics any different from the problem of the intersubjectivity of any other sector of knowledge? Or, to put it differently, is there any reason to think that communication by means of sentences belonging to the language of mathematical discourse diverges from communication by means of other sorts of sentences? A temptation to acknowledge an essential difference of this kind may be motivated by the following three observations. The first draws on a distinction between, on the one hand, objects that may be seen, smelled, touched, or generally perceived by the senses, and mathematical objects, on the other. In the case of a sentence about an object of the first kind, we frequently are able to resolve a perplexity about its meaning simply by pointing at an object of the appropriate sort. For instance, if it turns out that someone is perplexed about the expression 'green apple' as figuring in an utterance, we may show him something green as well as an apple. Now, if you assume that mathematical expressions denote mathematical objects, a similar perplexity is not always so easy to resolve. Obviously, in some cases, as for instance of number-words standing for finite numbers, the situation does not essentially diverge from the previous one. Given that you suspect a misunderstanding about the word 'two', you may indicate to your partner two pears, for instance. Although this method, fallible as it is, may be extended so as to be applicable to various kinds of mathematical expressions, it cannot cope with, say, expressions standing for infinite cardinal numbers. Thus, if one looks at mathematical communication from a perspective that assumes ostensive definitions, then doubt emerges about how we can communicate about the infinitistic part of mathematics. The view here discussed bears a close affinity to that of Hilbert. According to him, it is the nature of infinity that should be explained, since, first, it is less clear than other mathematical concepts, and second, arbitrary operations with this concept may be responsible for mathematical and logical paradoxes. (Hilbert 1926) The explanation of infinity, or rather, justification by finitary means of statements relating to infinity was one of the objectives of what is known as Hilbert's Program.

The second observation emphasizes, contrary to the above standpoint, that in mathematics we have a much better chance to communicate than in other subject matters. Mathematics is viewed here as a hard science and, perhaps together with some empirical sciences, is contrasted with subjects like philosophy, theology, and literature. An observation that lies behind this view can be summarized by saying that discussions among representatives of the latter disciplines sometimes give an impression that not only the participants are unable to reach any conclusions, but also that there is no guarantee that they are talking about the same thing and understand each other. Mathematicians or, say, physicists have, on the contrary, a last recourse in their controversies: proofs and experiments, respectively. Thus, in a controversy over the validity of a mathematical theorem, if they have a candidate for its proof, they can examine it step by step, and come to a conclusion about whether or not the proof is correct. Moreover, in order to make this procedure effective (in cases where it is possible), the mathematical theory to which the disputed theorem belongs should be subjected to formalization and the proofs examined—formal.

A hard-headed skeptic about the possibility of communication may voice an objection that nothing prohibits a mathematician from systematically misinterpreting a step in a proof, or simply ignoring an essential similarity between some strings of symbols. At that point, the third difference between the intersubjectivity of mathematics and that of other disciplines comes into play. In the case of a wide class of formal theories, the relation of being a proof of a formula is decidable. Accordingly, the question of whether or not a given syntactic structure is a proof of a given formula can be decided mechanically, so we may ignore the above skeptical objections.

What emerges from these considerations concerning the intersubjectivity of mathematics is praise of formalization. Accordingly, it is the view of a considerable group of mathematicians that a guarantee that mathematics is communicable should be sought in the formalization of the language of mathematical discourse. Moreover, this belief is independent of whether they accept a radically formalist conception of mathematics or not. This position was very clearly voiced by Stanisław Leśniewski. While writing about these matters in 1929, he began with a confession that he is far from believing that mathematical formulae are merely sequences of syntactically allowable signs devoid of any meaning. Similarly, he opposes the view that mathematical proofs are only mechanical operations on such sequences. On the contrary, he continues, the mathematical formulae are 'intuitively valid' and have meaning. Despite the last conviction, he nevertheless decides to undertake a formalization of mathematical language as he does not know of any other efficient way of communicating his mathematical intuitions to the reader. Now, seeing the matter in the light of the Skolem-Löwenheim theorem, these views appear far too optimistic. Formalization must take place in some language and usually is done in a first order language. Such a language, however, does not allow for differentiating standard and non-standard models. Thus, although some intuitions may be conveyed by displaying the details of a given formalization, to communicate fully with others in an efficient way requires us to indicate what model is intended, and this is precisely the thing we cannot do in the first-order language.[3] This objection does not apply to Leśniewski's view, however, since he is not committed to a first order theory.

One may find more contemporary opinions that a 'guarantee' of the intersubjectivity of mathematics consists in the decidability of relations obtaining between inscriptions of a formal language. As this view is very clearly voiced in Grzegorczyk's (1974) work, we will base our discussion on his exposition. As he explains, a set (or a relation) Z of inscriptions is *decidable* or *computable*, "if there exists an effective method which makes it possible to decide for any inscription whether it is, or is not in Z." (*ibid.*, p. 453) This means that one should view this guarantee as rather limited, since not all formalized theories allow for the decidability of such relations. Formalized mathematical theory defines what its sentential formulae and proofs are; this definition is framed only in terms of syntactic concepts. The relation that according to Grzegorczyk is essential for the issue of intersubjectivity is that of a given sequence of inscriptions being a proof of a formula. Given that this relation is decidable, he considers it a guarantee of intersubjectivity. (*ibid.*, p. 457) It may be seen as a last instance to which mathematicians who disagree over the validity or meaning of a mathematical statement can take recourse. That is, given that they have what they believe to be a proof of a disputed statement, they should submit it to a decision procedure that will show whether it really is so. If the apparent proof turns out to be mistaken, then it means that at least one of these mathematicians misunderstands the formal system. The misunderstanding can be traced back to a faulty grasp of a symbol or a syntactic rule of the system.

However, there are two rather evident objections to this proposal of what a guarantee of intersubjectivity is. First, its application is relative to a given formal theory: the procedure can only answer whether mathematicians, working in the same formal theory, understand a mathematical statement in the same way, as it is represented in this formal language. More importantly, due to incompleteness results, for some sentences the outcome of the procedure will be irreparably negative. Given an undecidable sentence, that is a sentence that has neither a proof nor a disproof in a given formal system, the procedure may only show that any alleged proof of this sentence is mistaken. To improve on this approach we can characterize intersubjectivity in terms of the decidability of the relation of being a proof of a formula in *some* formal systems that attempt to represent a non-formalized mathematical theory. To talk here of a *guarantee* of intersubjectivity, however, is highly misleading. Imagine a mathematical statement belonging to intuitive mathematics which is a point of controversy between mathematicians. What Grzegorczyk proposes boils down to a hope that in case of such a controversy, it will be possible to devise some formalizations of the fragment of intuitive mathematics, and in at least one of the resulting formal theories provability will be decidable and moreover the controversial statement will be represented by a decidable sentence. Now, given that all the above hopes are fulfilled, we have a criterion that decides, for any proposed formal proof, whether or not it is a proof of the sentence in question.

Nevertheless, in spite of the rather narrow area of application of Grzegorczyk's method, its merit consists in an attempted reduction of the much more complicated issue of intersubjectivity of mathematics to a possibility of identifying shapes of signs that are 'approximately the same'. As he stresses,

(...) an effective method of stating something about inscriptions must take into account the visually recognizable characteristics of the form of inscriptions. Thus, such a method is, in a sense, visual, since it consists in inspecting an inscription carefully and in transforming it graphically. (*ibid.*, p. 453)

Thus, the method of testing whether something is a formal proof of a given sentence amounts to the following directive: watch the expressions carefully and then perform some graphical transformations.

Let us finally reflect on whether a formalist-style explication of the intersubjectivity of mathematics is accessible to intuitionists. Both Brouwer and Heyting draw a radical distinction between mathematics and its language. First, they insist that it is the *language* of mathematics that is subjected to formalization, not mathematics itself. Second, Brouwer holds that any language, including that of mathematics, is an irreparably imperfect means of communication. Given these two claims, the early intuitionists can hardly see a 'guarantee' of intersubjectivity in the formalization of mathematics. And even if such a hostile attitude to formalization is moderated, as seems to be the case in Heyting's philosophy, the issue of the application of the above method is at best programmatic. For this to be possible, a crucial category of intuitionistic constructions, in terms of which the intuitionistic interpretation of logical connectives and quantifiers is given, should be characterized formally. As both Brouwer and Heyting opposed, on rather ideological grounds, this possibility, they had to appeal to a similarity of intuitions, or similarly understood basic mathematical notions. But if we are troubled by the issue of the intersubjectivity of mathematics, should we consider their reluctance to appeal to formalization a serious demerit of their conception of mathematics? We have been arguing that the 'guarantee' of intersubjectivity that Grzegorczyk identifies with the formalization and decidability of the relation of provability is not at all a guarantee since (1) the problem of non-standard models emerges, (2) provability is not always a decidable relation, (3) if a sentence is undecidable in a given formal system, the procedure may only show that any proposed formal proof of this sentence made in this system is not a proof of the sentence in question, and (4) the criterion is limited, as it only permits us to decide whether or not a *given* syntactic structure is a formal proof of a given formula. Thus, provided a number of additional premises are fulfilled, the formalist has a sort of last instance argument in controversies over the validity of mathematical statements. The intuitionist does not have anything similar to this. Nevertheless, the formalist's advantage over the intuitionist is on the verge of being illusory. Even if it is found that a given syntactic structure is a proof of the sentence in question, a dispute may break out over whether the formal system adequately captures an intuitive mathematical theory. One possible consequence of this dispute may be a controversy over whether a given formula of a formal system adequately represents the *meaning* of a given statement of intuitive mathematics. This sort of controversy is not easily resolved. Thus, the conclusion is that whenever mathematicians agree on a given formalization of some intuitive mathematical theory and moreover some additional premises are fulfilled, they have a last instance argument in disputes over validity and meaning. However, since any serious dispute over meaning will likely result in a dispute over allowable formalization, their criterion is of little help. Thus, the matter may be summed up by saying, paradoxically, that if we see that we agree on meaning because we agree on a given formalization of some part of mathematics, and moreover, some other facts obtain, then we will see that we have agreed on meaning. But if we are not so certain that we agree (and this may be displayed in our hesitancy to accept a given formalization), then the decidability of the relation of provability is of no help. These rather simple observations should remove the objection to intuitionism that draws on some apparently serious consequences of the intuitionists' refusal to formalize intuitionistic mathematics: It is illusory to see a guarantee of intersubjectivity in formalization.

We have now come to a more detailed investigation of various justifications of the intuitionistic revision of mathematics. We begin with Brouwer's philosophy.

2
BROUWER'S PHILOSOPHY

Philosophical remarks appear in many of Brouwer's papers from various phases of his career. One can find them in his doctoral dissertation (1907) as well as in his late Cambridge lectures, delivered between 1947 and 1951. With respect to many significant themes, it is rather striking how little he changed his conceptions. Even more, some passages, like those presenting his account of language and linguistic communication, the relation between logic and mathematics, or abilities of the knowing subject are repeated in his later works almost word for word, as if they were quotations from his earlier papers. That is not to say that all Brouwer's views remained unaltered. For instance, perhaps his most important concept, that of intuition, was substantially modified. Similarly, his mathematical notions evolved considerably. Nevertheless, the relative stability of Brouwer's philosophical views justifies a reconstruction of his whole philosophy, without paying much attention to distinctions between periods or phases of the development of his philosophical thought. Thus, we shall attempt an overall and somehow ahistorical presentation of his philosophy, and only exceptionally, if a concept under consideration went through a process of substantial changes, shall we remark on its origin and evolution.[4] To repeat, the objectives of our presentation of Brouwer's philosophy are to find out whether his arguments are strong enough to justify the proposed revision of mathematics, whether it was an account of meaning that moved him to question classical mathematics, and whether his conception of mathematics allows for intersubjectivity.

Embarking on the task of reconstructing Brouwer's philosophy, one soon encounters another methodological quandary. It may be suggested that Brouwer's opposition to classical mathematics was somehow motivated by his general outlook on life. For instance, one may speculate that his admiration for Schopenhauer's[5] philosophy may have motivated his rejection of objective reality as existing independently of whether or not human beings experience it. In a similar vein, it may be argued that passages of the Bhagavadgita, quoted by Brouwer several times,[6] may have something to do with his ideal of pure mathematics. A bit absurdly, his pessimistic views on his fellow-men, hatred of physicians and medications, and other personal likes and dislikes may also have been linked to his revision of mathematics. So, the question is, how far should one go in reconstructing Brouwer's philosophy? Should one try to take into account his interest in the Bhagavadgita, Schopenhauer's views and his personal likes and dislikes in the hope that his rationale for revising mathematics would somehow 'follow' from them? Or, quite to the contrary, should one ignore such interests and follow the feeling that in a reconstruction of somebody's philosophy of mathematics there is no place for his personal attitudes and interests? Brouwer's style of writing makes these question even more difficult to answer, as his papers often defy rules of scientific publication: for instance, passages from the Bhagavadgita neighbor with arguments for rejecting the principle of the excluded middle or with mathematical proofs. Nevertheless, we accept here a rough distinction between factors that motivate somebody into developing his philosophy and the philosophy itself. And although finding out what Brouwer's outlook on life and his perception of Schopenhauer and Eastern philosophies were may tell us a lot about his personality or even help us to understand his philosophical position, it has little bearing on the validity of his arguments for mathematical intuitionism. That is, while attempting a reconstruction of his philosophy of mathematics, we shall try not to invoke the issues mentioned.

The discussion of Brouwer's philosophical positions splits naturally into three parts. The first deals with the knowing subject and its 'functions': consciousness, mind, intuition, and a faculty that enables the creation of causal links. The second discusses the relevance of his concept of intuition for his vision of mathematics. The third concerns language and related phenomena: logic, as well as the latter's relations to mathematics.

1. THE KNOWING SUBJECT

In attempts to reconstruct Brouwer's thought in a way that shows how his intuitionism follows from, or was motivated by, his philosophical positions, one is naturally tempted to take a closer look at his epistemology. The move is tempting because his answers to questions like 'What can be known?' or 'What is to be known?' may somehow lead to the decision about what mathematical objects are allowable or what a correct concept of truth is. So, giving in to this temptation, we shall concentrate first on his epistemology. To present Brouwer's epistemological views one needs first to sketch out a 'structure' of the abilities which he assumes, the knowing subject has. Unfortunately, right at the beginning of this enterprise we encounter the most obscure notion of his philosophy, namely the notion of *consciousness*.

1.1. *Mind, Time and Objects*

The salient feature of consciousness, as Brouwer says, is that it experiences sensations. However, they appear in a way that seems to be more proper to dreaming or being half-asleep than to being fully awake. As Brouwer puts it, "Consciousness in its deepest home seems to oscillate slowly, will-lessly and reversibly between stillness and sensation." (1948, p. 480) In this phase there is no distinction between the subject and experienced sensations. Accordingly, sensations do not point to external and independent objects that are distinct from the consciousness. In Brouwer's words, the process of 'objectification', which is responsible for the emergence of the 'intentional' character of sensations, has not started yet. That the consciousness oscillates between stillness and a flow of sensations, and that sensations are passing and being replaced by others, is possible due to time, or more precisely, to what he calls 'a move of time'.[7] In moving time, one sensation gives way to another. The former has already passed, the latter is just 'now'. That is how, according to Brouwer, the distinction between past and present is born. Time is associated in Brouwer's system with the emergence of another faculty, which he calls the *temporal attention*. He defines this as "(...) the perception of a *move of time*, which is the falling apart of a life moment into two distinct things, one of which gives way to the other, but is retained by memory." (1954, p. 523) With the emergence of the temporal attention, the reversibility of the flow of sensations is gone. They become ordered according to whether they occurred earlier or later. One may thus say that the experience of inner time permits the ordering of meandering sensations. On the other hand, once the sensations become ordered, the flow of sensations is sundered from that which experiences them. Sensations appear to be independent of the subject, the appearance being that they originate in 'objective reality'. The consciousness, now fully separated from past and present sensations, becomes the *mind*. Its role is to receive an ordered flow of sensations. It is perhaps worth mentioning here that for Brouwer the described movement, from the consciousness, via temporal attention, to the mind is a degeneration. Namely, he considers the consciousness the source of ultimate happiness and peace, the departure from which is the original human sin. Yet needless to say, and here Brouwer agrees, had the mind not evolved from the much more superior consciousness, human beings would not be able to practice mathematics and science, invent technology, and engage in social life.

It is worth stressing that 'a move of time' is for Brouwer the source of all cognitive activities of the knowing subject. On the one hand, "the falling apart of a life moment into two distinct things, one of which gives way to the other, but is retained by memory" (Brouwer 1981, p. 4) gives rise to the mathematical intuition of two-ity, which we shall discuss later. On the other, as it leads to a faculty Brouwer calls 'causal attention', it allows the subject to order sensations and their complexes into causal links and objects.

The mind experiences complexes of sensations and in their flow it identifies 'similar' sequences of sensations, that is, sequences of a certain stability, such that their initial fragments are usually followed by the same series of sensations. In his dissertation (1907) Brouwer is more explicit about the mechanism by which causal chains emerge. According to him, the active

cess that leads to the emergence of that sphere mostly consists in postulating new entities, objects, properties but also sensations which the subject does not experience. This is, according to Brouwer, an effect of projecting mathematical structures on an array of sensations, whose paradigmatic example is the ordering of sensations into sequences. In this process some elements (or parts) of the mathematical structure may remain 'empty'. This means that they do not refer to a sensation or a complex of sensations that the subject experiences. Nevertheless, the subject postulates an 'invisible' sensation (or a complex) that is correlated with such an empty mathematical structure. The existence of such postulated entities: objects, properties, causal chains and links and 'invisible' sensations, is a 'physical hypothesis'. The process also has another side as complexes of sensations that spoil the projected regularity are neglected. That is, the mind suppresses their presence and does not regard them as pointing to an object or a property. (1979, p. 396; 1929, p. 418; 1948, p. 481)

As Brouwer notes, the causal attention often yields failures and disillusion. The mind constantly meets resistance from sensations and complexes of sensations that do not easily submit to being put into regular and stable mathematical forms. Also, a belief in causal connections obtaining in nature is frequently shattered by irregularities among sequences of sensations. For instance, it happens that by joining two such causal sequences where the last element of the first sequence is the same as the first element of the second, the sequence that emerges is not causal. (1948, p. 481) But in spite of the fact that the 'ordering' of sensations into causal chains gives rise to many failures and mistakes, it is indispensable for all practical and scientific human enterprises, at least as long as the ultimate human goal is to subordinate nature to human needs—claims Brouwer. He believes, however, that all sciences with the exception of mathematics serve this end. As they allow a man to predict the future, they enable him to bring nature under human control, at least to a certain extent. And sciences stem from the causal attention of the knowing subject, as do all the practical human interests. In describing how the sciences work, Brouwer points to a mechanism similar to the process of arranging of sensations, in which the mind projects mathematical structures on complexes of sensations. The order of phenomena, as introduced by this process, is merely a heuristic hypothesis: it permits us to utilize more easily the plenitude of data to achieve practical goals. One can hardly talk of laws of nature that are true and independent of the knowing subject. Similarly, truth is not a feature of scientific theories. Brouwer is convinced that practical utility of such theories does not testify to their truth.

Now, we may ask with Brouwer why one is able to comprehend a plethora of phenomena and aggregates of sensations stemming from our various senses in the form of general mathematical laws. (Brouwer 1979, p. 399) Why can one predict future sensations or phenomena with the help of laws of nature, whatever the status of the latter may be? In an attempt to answer this query,

Brouwer maintains that laws of nature are 'constructed' by scientists and that the technology of scientific devices mediates actively in this process. He starts with the brief remark that such devices are built from more or less the same material and according to similar design. This remark should obviously be understood in the light of Brouwer's conception of objectivity: 'similar bodies' means that the sequences of sensations that point to one and to the other are similar. In most cases the sensations which scientists attempt to arrange in regular patterns come *via* the devices mentioned. That is, the scientist tries to impose a mathematical structure on sensations which are obtained with the help of a scientific apparatus. And now, Brouwer claims, the sensations and the objects they point to look regular *because* the scientific devices are regular and alike. Thus, in the case of astronomy, he draws the following conclusion:

(...) the laws of astronomy are no more than the laws of our measuring instruments when used to follow the course of heavenly bodies. (*ibid.*, p. 399)

1.2. *Brouwer and the Problem of Other Minds*

Brouwer's views, as presented above, may lead to the objection that his philosophy prohibits the existence of other minds. This objection has a crucial bearing on our project since if one holds such a position, then clearly the question of what makes knowledge communicable in his system has no sense. Moreover, if the objection is justified, one could only treat Brouwer's views as an anecdote, and perhaps only wonder how he managed to combine his professed views with everyday activities as a teacher, lecturer, and author of mathematical publications.[8] Similarly, Brouwer's arguments for the intuitionistic revision of mathematics would be undermined if they drew on solipsism.

What gives rise to the accusation of solipsism is a disturbing argument in Brouwer's (1948) paper against 'the plurality of minds'. Before we analyze it, let us recollect the most general characteristics of the Brouwerian mind. Firstly, the mind separates itself from the consciousness in the same process in which sensations receive an independent status. Secondly, apart from being a receiver of passing sensations, it is a free agent with will and emotions. And thirdly, the mind has several faculties, like causal attention, temporal attention, and mathematical intuition, which permit it to arrange sensations in regular patterns, postulate an objective reality behind them, develop science and mathematics, and engage in practical tasks. This characterization of mind leads to a considerable shift in the meaning of the question about the plurality of minds. The problem of other minds is reduced to a query about whether, by using the faculty of causal attention, the mind of the subject may ascribe minds to some specific complexes of sensations that point to special objects, that is, to human bodies. Let us see how far one can go in ascribing minds to such postulated objects. To begin, by a human body one should understand in

Brouwer's framework a specific complex of sensations, or perhaps more accurately, a postulated object to which the complex points. The subject, which must be 'I' in this case, observes that certain sensations and their complexes depend on his will. On the other hand, I observe that I cannot exert any influence on some other sensations, as I cannot make them disappear, arise, or change their content. Some such 'resistant' sensations seem to be grouped together and the complexes they form seem to be similar to complexes referring to my body. What is more and most crucial to Brouwer's solution of the problem, while concentrating on some sensations that are 'associated' with complexes pointing to other bodies, I recognize that their arrangement is similar to (or even the same as) the structure of sensations which I naturally consider to be brought about by the working of my will. (1948, p. 484) Speaking in a more objective manner, I recognize that certain bodies act, perform, and behave in the way I do, and by no means are these activities outcomes of my will. At this point, however, one meets a lurking paradox. In Brouwer's thought whatever exhibits a regularity is a creation of the mind, after all, and so the mind that recognizes a similarity of some bodies seems to be discovering a fact about its own functioning. To moderate the appearance of oddity, it is perhaps worth recollecting that although the regularities of sensations are created by the mind, they are brought about in an unconscious and spontaneous way. So, as an act of recognizing is in most cases a conscious process, one may avoid the lurking paradox by claiming that in recognizing, say, a similarity between two sensuous sequences, the subject becomes aware that two sequences in which it unconsciously and spontaneously ordered some sensations are similar. To come back to Brouwer's account, the recognition of sensations that point to activities of other human bodies, the activities that the observer has no power to control, leads to a postulate that other bodies are agents of their activities. It is assumed that the individuals perform various acts whose causes lie somehow 'in' the bodies. Thus, the subject postulates that the other bodies contain an active factor.

Can one, however, proceed further and, by following the causal attention, ascribe mind to other bodies? The answer must be in the negative, Brouwer argues. In his system, if the causal attention ascribed mind to another body, it would mean that it is capable of experiencing other minds as (clusters of) sensations. The concept of the mind experiencing other minds, which he dubs a *mind of the second order*, seems odd, however. Moreover, he attempts to prove that if one assumes that minds can be ascribed to other bodies, one is forced to accept the existence of a mind of an arbitrary high order. Let us take a closer look at this disturbing 'proof' of Brouwer's.

Suppose the subject-mind ascribes minds $M_1, M_2, M_3, \ldots, M_n$ to individuals $I_1, I_2, I_3, \ldots, I_n$. This amounts to the claim that the subject-mind has an insight into minds associated with individuals $I_1, I_2, I_3, \ldots, I_n$. As Brouwer assumes that the vital role in gaining this insight is played by the causal attention, which arranges sensations, the knowledge of other minds boils down to experiencing some sensations or complexes of them. Thus, Brouwer says, the subject mind 'elevates itself' to a mind of second order that experiences as sensations minds of the first order i.e. $M_1, M_2, M_3, \ldots, M_n$. Proceeding consistently, however, we need also to assume the plurality of minds of the second order. That is, we should assign a mind of the second order to any individual $I_1, I_2, I_3, \ldots, I_n$. Otherwise, the status of the subject mind (of the second order) would be distinguished. So, we now have a totality of the minds $M_{k,l}$, all of them of the second order. As the notation suggests, a mind $M_{k,l}$ is ascribed to the individual I_k and it receives sensations by which the mind M_l presents itself. In Brouwer's framework, talk about the plurality of minds of the second order is meaningful only if the subject-mind could ascribe such minds to some bodies. This would mean that the subject-mind experiences minds of the second order. That is, it assigns such minds to individuals. This can only happen if it is elevated to the status of a mind of the third order. A mind of the third order experiences sensations by which a mind of the second order is given. Again, proceeding consequently, we need to postulate a totality of minds of the third order, as all the minds should be on a similar footing. Thus, one needs to introduce an aggregate of minds $M_{k,l,n}$ where k refers to the individual to which a third-order mind is being ascribed, while l and n fix which mind of the second order is experienced. Thus, Brouwer suggests that once we ascribe minds to other individuals, we are in for a *regressus ad infinitum*. There is no definite order which should end the hierarchy of minds of an increasing order. (1948, p. 485) This regress to infinity is one reason for his rejection of the idea that, in the causal attention, minds of other individuals are given to the subject. But whether his argument is correct depends on establishing an additional premise. It should be shown that minds of a different order must be different. Otherwise, we may simply claim that mind M_k of the first order is identical with mind $M_{k,l}$ of the second and so on, and the above regress will not arise. Moreover, as this is a view one would naturally assume, an argument is urgently needed. Another crucial problem is what in Brouwer's framework can serve as a principle of differentiating between minds. The way the concept of mind is defined suggests that such a principle of individuation can only be stated in terms of differences between sets of experienced sensations. However, a straightforward stipulation that minds are different only if they experience different sensations will not do, since it will split the subject mind into indefinitely many minds. Thus, the principle must rather invoke a distinction between essentially different sensations or clusters of them, such that sensations arising from my body and environment are essentially different from sensations by which the mind of the other individual is given to me. However, to assume such a distinction is question-begging since it blocks, without any reasons, the identification of the subject-mind with minds of higher orders.

Thus, Brouwer's main argument is not convincing. Nevertheless, the reader may also suspect still another ground for his rejecting the idea that the subject may ascribe minds to other individuals, which is the lack of any clear criterion for deciding to which individuals such an ascription can be made. As he puts it, to be consistent one should attribute minds to animals as well as to humans, or more precisely, to sensuous complexes pointing to animal bodies as well as to those referring to human ones. (*ibid.*, p. 485)

The discussion about the plurality of minds is closely linked to assumption of an objective world, that exists independently of whether or not anybody experiences it. In Brouwer's framework, the only sense that can be given to this claim is to understand it as concerning the existence of a special subject. Thus he writes about

(...) a collective super-subject experiencing an objective world which exists independently of the supposed human subjects that appear and disappear in it, which remains when all supposed human subjects have vanish, and would be, even if there had never been human subjects called into existence. (*ibid.*)

One may wonder why the assumption of the existence of the objective world translates into the positing of such a super-subject. The implicit but crucial premise of this reasoning is his rejection of 'absolute' truth: any claim about a mind of a certain order, to be meaningful, must be transformed into a statement about the experiences of some subject; similarly a claim concerning the objective world invokes a question concerning the experiences of some subject. As we will see, rejection of the concept of bivalent, verification-transcendent truth will play a crucial role in his arguments against classical mathematics and logic. This concept licenses the classical mathematician's belief that a statement may be true, no matter whether its proof can or cannot be known. In a similar vein, the classical mathematician may believe a statement to be false, although it may happen that nobody can show how a contradiction follows from its assumption. For the moment, however, we shall leave Brouwer's rationale for repudiating bivalent truth and concentrate on what is claimed in his proof of 'no plurality of mind'.

While contemplating the proof, one finds first that although the causal attention is incapable of attributing minds to other bodies, it nevertheless may ascribe to them some active factor or the will. Moreover, in the same paper we find a statement that certain behavior of human individuals may be attributed to their reason, but not to their minds. Thus, one is tempted to see in this argument a point about how to use the term 'mind' in Brouwer's system. The issue is, however, far from clear, as no explanation of his other concept, 'reason', is given. Nevertheless, the way he poses the problem suggests that one should understand his position as merely methodological. That is, he seems to be saying that there is no hope, if one relies on the causal attention only, of attributing minds to other human bodies, although one can treat them as human beings capable of voluntary acts and equipped with reason. This stance is extended further to sciences that allegedly deal with the mind. As they stem from the activities of the causal attention, Brouwer consequently holds that scientific knowledge of conscious and mental phenomena is impossible. He recommends that psychology should study man and animals alike, as "automatic living organisms without mind and without free will." (*ibid.*, p. 485) On the other hand, however, in his (1933) paper he concedes while discussing the process of imposing an order on sensations that

A very essential *hypothesis* in the mathematical view of one's fellow-men e.g. is the hypothesis that there is in each of them a mathematical-scientific mechanism of viewing, acting and reflecting similar to ones own. (1933, in VS p. 420; my emphasis)

As the mechanism mentioned is characteristic of the mind, the hypothesis says that others *have* minds. However, as we shall see, to say in Brouwer's system that something is a hypothesis amounts to denying that truth applies to it.

Brouwer's criticism of the inability of the causal activity to introduce other minds also has some moral overtones. This faculty, according to him, treats everything that the subject experiences as elements of a causal net constructed by the subject. Because of that, various objects, properties, causal chains, and sensations are viewed by the subject merely as 'means' by which a desired states of affairs ('aims') can be achieved. In other words, the subject that is causally directed views reality only as a plenitude of variously linked objects, properties, and sensations whose only role is to serve the subject's needs. Thus the issue of solipsism is linked to another query, namely the possibility of 'turning off' the subject's causal attention. One may also ask if there are other faculties which allows the subject to experience his fellow men. As to the first query, Brouwer believes one can revert to one's initial state, that is, again to become the pure consciousness. If that process is to succeed, the distinction between someone's own 'intimate' sensations and sensations 'foreign' to the subject should melt away. The removal of the distinction and the return to the status of consciousness is possible, according to the Dutch mathematician, in the arts (especially music), in mystical revelation and in free mathematical creativity. As to the second question, we find a remark to the effect that a subject who 'puts in brackets' his causal activity is capable of experiencing other humans. As Brouwer says, "only through the sensations of the other's soul sometimes a deeper approach is experienced." (1948, p. 485) One may interpret this remark as an assertion that there is some non-causal experience of other human beings, which is made possible by emotions, passions, and feelings of personal qualities by which the subject attends to other human beings. As they are almost always externally directed, they presuppose the existence of what he calls other 'souls'. From a common-sense perspective, however, it is unclear how the subject, whose various data do not split into 'intimately own' and 'foreign' sensations, may experience anything different from itself, as something at which emotional feelings are directed. However deep this tension may be, such an interpretation of Brouwer's solution to the problem of other minds allows one to reconcile his metaphysics with his other views, like those about human collaboration and language, and with his everyday conduct as a lecturer and a writer.

To sum up the discussion of Brouwer's 'no plurality of mind' argument, its thesis is seen here as methodological: what he calls 'mind' cannot be ascribed by the causal attention to other individuals. On the other hand, this faculty may attribute free will to other bodies. Moreover, certain behaviors of other individuals can be viewed as traceable to their 'reason'. Also, once the causal attention is 'turned off', the subject receives a deeper understanding of his fellow men. Finally, he seems to agree that it is an essential hypothesis that others have minds. In the light of the foregoing, we are inclined to conclude that accusing Brouwer of solipsism is unjustified and results from a selective reading of his remarks about other minds. Saying this we do not mean that we have an interpretation that makes full sense of these 'thorny' and conflicting passages of Brouwer's writings.

1.3. *The Basic Intuition of Two-ity*

It is a part of Brouwer's doctrine that the mind evolved from the consciousness when, due to the flow of time, coming and passing sensations were separated from it and the experiencing agent, i.e. the mind emerged. Apart from such abilities of the mind as the temporal attention or the causal attention, Brouwer assumes still another faculty, dubbed 'mathematical abstraction', that permits the removal of all sensuous content from any two consecutive sensations. This is how one comes to a pure form of succession. As he puts it, intuitionistic mathematics has

(...) its origin in the perception of a *move of time*, i.e. of the falling apart of a life moment into two distinct things, one of which gives way to the other, but is retained by memory. If the two-ity thus born is divested of all quality, there remains the *empty form of the common substratum of all two-ities*. It is this common substratum, this empty form, which is the *basic intuition of mathematics*. (1952, p. 510)[9]

The way the intuition is introduced suggests that the concepts of the mind, the causal attention, and the mathematical intuition lie very closely to each other in Brouwer's philosophy. The intuition is a form of any succession of two sensations and, given that the mind is characterized as a receiver of temporarily ordered sensations, the intuition is indispensable to it. In mathematics it is required that mathematical knowledge be based on the intuition of two-ity. This demand delineates which mathematical constructions and proofs are acceptable in mathematical intuitionism. The 'unfolding' of the intuition makes possible the introduction of natural numbers, the construction of rationals and real numbers.

Before we proceed further, it is of some importance to mention that in his earliest writings (e.g. his doctoral dissertation) Brouwer accepted a somewhat different notion of intuition. In the later concept, the emphasis is on the succession of sensation: intuition is identified with 'the form' of passage from one sensation to the other. On the other hand, his dissertation characterizes intuition as

(...) the substratum, divested of all quality, of any perception of change, a unity of continuity and discreteness, a possibility of thinking together several entities, connected by a 'between', which is never exhausted by the insertion of new entities. (1907, p. 17)

He calls this intuition: 'the intuition of time'. Continuity and discreteness are characterized as 'inseparable complements', as if they both were primitive notions, neither of which was capable of being constructed from the other. Clearly, at this time Brouwer held the concept of continuum to be primary and undefinable. Although in the above quotation it is assumed that continuity is united with discreteness, both notions were separate in the constructivist mathematics of that period. There was a considerable tension between theories of real numbers on the one hand and methods accepted in mathematics of natural (rational) numbers, on the other. Brouwer was preoccupied with the task of introducing the 'totality' of real numbers that would use only the resources of constructivist mathematics, which we can think of as more or less the resources of a theory of natural numbers. Herman Weyl also aimed at a similar goal before 1918. But, as it was not known how to solve this problem, Brouwer had to make a concession and treat discreteness and continuity as two irreducible, undefinable and primary notions. In the early twenties, however, Brouwer made progress in solving the mathematical problem. With the introduction of the intuitionistic concept of set (called *species*) and by generalizing the notion of numerical sequence (dubbed *spreads*), Brouwer introduced the species of real numbers by showing that it is equivalent to the species of spreads of rational numbers. Consequently, the continuum disappeared from his account of intuition, as the succession of elements turned out to be prior to continuity. As some commentators suggest (e.g. Dąmbska 1976), it is of some historical importance to notice that Brouwer's later concept of intuition as the form of all two-ities resembles Kant's pre-critical account of a faculty with the help of which the subject 'synthesizes' composite objects. In *De mundi sensibilis atque intelligibilis forma et principiis*,[10] while talking about 'representing the notion in the concrete', Kant assumes this to be possible due to a temporal condition, as it enables the subject to arrive at the concept of a composite by the successive addition of part to part. The condition of successive addition is a close counterpart of the Brouwerian 'form' of passage from one sensation to another. Later, in the *Critique*, Kant called the faculty of representing notions in the concrete 'pure form', and identified it with continuous time. Thus, if this suggestion is correct, both of Brouwer's concepts of intuition resemble those introduced by Kant, although in the careers of the two thinkers they were entertained in reverse order.[11]

2. MATHEMATICS AND INTUITION

2.1. *Infinitely Proceeding Sequences, Spreads and Species*

What roles does Brouwer give to the intuition of two-ity? On the one hand, it is a means of introducing various mathematical structures, so it is responsible for the building and cogency of the 'new' mathematics. On the other hand, it allows the intuitionist to dismiss as empty those mathematical definitions and proofs that do not correspond to constructions accessible by following the intuition of two-ity. Let us first concentrate on the positive side and see how the new, 'intuitive' mathematics is to be built. The answer to this question is contained in two postulates which Brouwer calls 'acts of intuitionism'. About the first he says:

(...) the first act of intuitionism completely separates mathematics from mathematical language, in particular from the phenomena of language which are described by theoretical logic, and recognizes that intuitionist mathematics is an essentially languageless activity of the mind having its origin in the perception of a *move of time*, i.e. of the falling apart of a life moment into two distinct things, one of which gives way to the other, but is retained by memory. If the two-ity thus born is divested of all quality, there remains the *empty form of the common substratum of all two-ities*. It is this common substratum, this empty form which is the *basic intuition of mathematics*. (1952, p. 510)

In a later paper he adds:

This empty two-ity and the two unities of which it is composed, constitute the *basic mathematical systems*. And the basic operation of mathematical construction is the *mental creation of the two-ity of two mathematical systems previously acquired*, and the consideration of this two-ity as a new mathematical system. (1954, p. 523)

As the memory allows the subject to retain systems previously acquired, the repeated operation of creating a two-ity leads to constructions of arbitrary natural numbers or finite sequences of natural numbers. On the other hand, the possibility of thinking of this operation as repeating without end allows one to introduce the *infinitely proceeding sequence* of natural numbers (called *ips*). (*ibid.*) While this 'first act' describes what the intuition is, tells how basic mathematical structures are introduced, and draws a sharp distinction between intuitionistic mathematics and the language of mathematics, the second act is explicit about what sort of mathematical objects the intuition allows and how these objects are created. It recognizes two possibilities of generating new mathematical entities: infinitely proceeding sequences and species.

Firstly, a new mathematical entity may be introduced as

(...) *infinitely proceeding sequence* $p_1, p_2, \ldots$, whose terms are chosen more or less freely from mathematical entities previously acquired; in such a way that the freedom of choice existing perhaps for the first element p_1 may be subjected to a lasting restriction at some following p_n, and again and again to sharper lasting restrictions or even abolition at further subsequent p_ns while all these restricting interventions, as well as the choices of p_ns themselves, at any stage may be made to depend on possible future mathematical experiences of the creating subject (...). (1952, p. 511)

A natural way of viewing the choice sequence is to assume that it is generated by a process, divided into stages, such that at any given stage k the k-th element of the sequence is introduced, together with a restriction on the values of its future elements, from k-th on. As the choice of an element does not necessarily go together with the introduction of a restriction on its successors, we may conventionally assume a restriction which is satisfied trivially by all elements and equate 'no restriction' with the imposition of a trivial one. For instance, in the case of a sequence of natural numbers, a trivial restriction may simply require any element of the sequence to be a natural number. With this convention we may represent the choice sequence as a sequence of pairs: $\langle \alpha(1), R_1^\alpha \rangle, \langle \alpha(2), R_2^\alpha \rangle, \langle \alpha(3), R_3^\alpha \rangle, \ldots$ where $\alpha(k)$ is the k-th element of the sequence and R_k^α is a restriction imposed at the k-th stage. To give an example of a choice sequence, let us imagine a prescription that says the elements $\alpha(1), \alpha(2), \alpha(3), \alpha(4)$ can take the value of any natural number, then at the 5-th stage the requirement is imposed that from that stage on elements be smaller than 137, further at the 10-th stage it is postulated that the following elements be even. The talk about a prescription is somehow misleading here as one needs to keep in mind that the restriction is not accepted until an appropriate stage. That is, the subject who carries out the generation does not know in advance, or at an earlier stage, which restriction will be imposed at any later stage. This supports the crucial idea of intuitionism that in reasoning about these sequences one may rely only on information gathered from an inspection of the processes of introducing *finite* segments of choice sequences. This idea lies behind many intuitionistic proofs in the theory of choice sequences and the theory of real numbers. Depending on the restrictions, the resulting sequence can be more or less determinate. Thus, at the one extreme, if the first restriction boil down to a function determining the value of any element of the sequence, we have a law-like (or pre-determinate, as Brouwer called it) sequence. On the other hand, if no restrictions are assumed, or in our convention, only trivial restrictions are introduced, the sequence proceeds completely freely. Such a sequence is called 'lawless'.

Moreover, imposed restrictions may lead to a 'break' in the process of generation of a sequence, since it may happen that at some stage k no number can satisfy the already imposed restrictions. At some point in his career Brouwer extended further this basic concept of choice sequence by assuming the possibility of accepting restrictions on restrictions, or as they are sometimes called, restrictions of the second order. For instance, one may imagine, that at the k-th stage of the generation of a sequence in which the already imposed restrictions (of the first order) require that all the following elements should be greater then 7, it is further required that no future restrictions can demand that the elements be even. However, given that our example correctly represents Brouwer's intentions, it is hard to see how one can find out whether restrictions of the first order satisfy those of the second order, and accordingly, how it may be decided whether at a given stage the sequence stops or proceeds further. As to restrictions of the first order, we may conceive of them as extensional properties of natural numbers. Contrary to that, since restrictions of the higher order operate on restrictions of the first order, there is good reason to think that they are intensional. To return to our example of the second order restriction, which requires that no restrictions of the first order can demand that the elements be even, it is not clear that the subject is always able to recognize whether or not a new restriction of the first order taken together with the earlier ones does not demand that the elements be *even*. It is too much of an idealization to assume that the subject conceives of all consequences of accepted restrictions of the first order. It may well happen that it is not known if from these restrictions the demand to accept only even numbers follows. Accordingly, what should the subject do, if in such a state of uncertainty the next generated element is even? Should this element be accepted and the sequence generated further, or should the decision rather be postponed until it is found out whether the first order restrictions do not conflict with that of the second order? However, it may well happen that it is not humanly possible to recognize whether such a conflict obtains. One may obviously have recourse to a syntactic characterization of restrictions of the second order. For instance, a restriction of the second order could require that in restrictions of the first order, as formulated in a given language, such and such expressions should not occur. However, this proposal will hardly match the overall anti-formalistic stance of intuitionism and moreover leads to doubt as to what the higher order restrictions are for. It is likely that such difficulties inherent in the concept of second order restrictions made Brouwer give up that notion, as he finally settled on the simpler concept of choice sequence and maintained that admission of higher order restrictions is superfluous and leads to unnecessary complications.[12]

Although much work has subsequently been done in the theory of choice sequences that went both in the direction of formalizing this theory and analyzing Brouwer's proofs, there are still some lingering doubts, mostly of a philosophical nature. Firstly and most importantly, it is problematic how one should construe the notion of free choice. On Brouwer's account free choices are effected by the knowing subject. This at once lands us in a rather peculiar doctrine of free-will according to which the subject is essentially free in choosing natural numbers. But, as the subject cannot easily be identified with a flesh-and-blood mathematician, it is not at all clear whether a faculty of choosing freely may be ascribed to humans. Moreover, if we leave aside the abstract knowing subject and its problematic relation to real people, and moreover realize that in selecting numbers from some set we may in actuality be biased towards some of them, prefer some and tend to ignore others, the question arises where in the real word an analogue of free choice can be looked for. It naturally springs to mind that such freedom is displayed by phenomena exhibiting physical randomness, whose paradigmatic example is casting a die. However, there is evidence that Brouwer rejected the idea that the free character of choice sequences can be identified or guaranteed by physical randomness, for instance that such a sequence may be generated by casting a die.[13] As to his rationale for the repudiation of physical randomness, we are again left only with speculation. Perhaps he feared that some imperceptible physical laws could introduce unwanted restrictions (regularities) that could vitiate the construction of the continuum. If this was so, he could hardly be convinced by an improved argument that makes an appeal to phenomena described by a truly probabilistic theory. A proponent of such an argument will probably agree with Brouwer that casting a macroscopic die is not a good guide towards genuinely free choices, as the process is likely governed by deterministic laws. Nevertheless, he will argue that with some ingenuity it should be possible to construct a *quantum die*, whose 'casting' yields results that depend in only a probabilistic sense on the initial conditions of the system. But for someone who is afraid that some hidden laws can introduce undesirable regularities and spoil genuine freedom, this move has only a very limited persuasive force. Finally, in an attempt to resolve the perplexity, one may turn to mathematical algorithms that generate random numbers, as applied for instance in computer programs. But obviously these are algorithms and the generated numbers are only quasi-random. Thus, as attempts to arrive at a more down-to-earth concept of free choice are not successful, we are left simply with a *postulate* that an ideal mathematician or Brouwerian subject has the capacity freely to generate natural numbers.

Secondly, restrictions imposed during generation of a sequence may turn out to be of a conflicting nature, which would make the process of generation come to a 'stop'. There does not seem to be a guarantee that the process of generating choice sequences will progress. Even if by introspection of the processes of generating initial segments of sequences α and β it turns out that the segments are identical and both sets of restrictions boil down to the same law, one could not conclude that the sequences are extensionally identical, since some future restriction may stop one, or the other, or both of them. One way out of this difficulty is to assume a straightforward stipulation to the effect that the sequence never ends. This stipulation, however, resembles closely the rejected notion of the second order restriction, as it amounts to a demand that future restrictions on the generation of the sequence should not contradict each other. Much more importantly, it is hard to see how this postulate can work in the intuitionistic setting. In effect, it should say that only such restrictions can be allowed as are possible to fulfill. However, it may happen that it is not known whether or not a set of restrictions is possible to fulfill. For instance, we can easily imagine that as an effect of earlier restrictions, some element $\alpha(k)$ is determined in the following way:

$$R(k) = \exists_{x,y,x,n}(\alpha(k) = 2^n \times 3^x \times 5^y \times 7^z \wedge n > 2 \wedge x^n + y^n = z^n).$$

Now, in the case of any given number we have a procedure that decides whether or not it satisfies the above condition. First, factorize that number and find if it can be represented as $2^n \times 3^x \times 5^y \times 7^z$ and next check whether $n > 2$ and x, y, z and n fulfill the above equation. However, despite this procedure, the subject may never find a number that satisfies this restriction! Moreover, given that Fermat's Last Theorem is true, $\alpha(k)$ *cannot* be found. Thus, we need some clarifications. First, what happens to the generated sequence if at a stage, say k, no number is known which satisfies the restriction on $\alpha(k)$? To put the matter jokingly, how long should the subject work on finding that number? Should the generation be broken after a number of unsuccessful tries, or rather postponed until some later phase in which either the number is found or it is proved that no number can satisfy the condition? The former solution makes breaking the generation of the sequence an empirical matter. One may say: you stopped developing the sequence because you did not try hard enough; but if you had made some more calculations, you would have found that number and could have generated the sequence further. On the other hand, the latter option is far too optimistic, as the problem may be truly undecidable. The example with Fermat's Theorem shows that, even if the condition imposed by the restrictions is decidable, there may be no effective procedure for finding the number that satisfies it. So the question is, What restrictions should be allowed for the subject to be able effectively to find numbers satisfying them?[14] The above queries concerning the concept of choice sequence will not be discussed further, as they lie somewhat outside the topic of the book.[15] Instead, while later proving a strong intuitionistic counterexample to the principle of the excluded middle, we present an argument that involves choice sequences. This may give more insight into how to reason about choice sequences than a lengthy philosophical discussion without the flesh of real intuitionistic mathematics.

The concept of choice sequence is essential for introducing the notion of intuitionistic *spread*, which undoubtedly is one of the most important concepts of intuitionistic mathematics, since constructing an appropriate spread is virtually the only method of generating structures of higher complexity than infinitely proceeding sequences. For instance, it allows the intuitionist to introduce an analogue of the classical continuum of real numbers. To come to terms with this concept, it is instructive first to present an illustration that Brouwer gave.[16] Imagine two people, A and B, of whom the first freely chooses a natural number and then announces it to his partner. Person B has a stock of symbols or 'figures', and given that A announces a number, he takes, in accord with some rule, one of three courses of action: First, he may put down a figure and tell his partner to choose another number; secondly, he may stop (or write down 'nothing') and not put down any more symbols, no matter what his partner later does; thirdly, he may stop and destroy all the figures he has written—this last option is usually referred to as *sterilization*. The first option is needed to generate infinitely proceeding sequences; the second guarantees the creation of finite sequences; the third represents the possibility of the destruction of a sequence in whose generation a contradiction occurred at some stage.

Now, in defining a spread we follow Brouwer's late account, as given in his Cambridge lectures (Brouwer 1981, pp. 13–15). One needs first to consider finite sequences of natural numbers, called *nodes*. Natural numbers that are elements of a node are called its *indices*. If a node consists of n indices, it is said to be of the n-th order. If the initial segment of a node coincides with another node, the former is called a *descendant* of the latter. For instance, the node $[1, 4, 7, 2, 9]$ is a descendant of the node $[1, 4, 7]$. What is important about these structures is that they can be enumerated by observing two principles: each node proceeds before all its descendants and nodes of a given order but differing only in their last indices are ordered in accord with the natural ordering of their last indices, the result being a partial ordering of nodes.[17]

The second step in introducing a spread consists in ascribing, in accord with a law, other mathematical objects, symbols, or as Brouwer has it, 'figures' to some elements of the species of nodes. Thus, it is said that a *spread* is a law that for any node either

1. yields a 'figure',
2. or yields nothing,
3. or sterilizes the process of generation and brings about the abolishment of the resulting finite sequence of 'figures'.

The law should satisfy the condition that for any non-sterilized node at least one non-sterilized immediate descendant can be found. Also, all descendants of a sterilized node should likewise be sterilized. The law should be effective and 'figures' only law-like, that is they cannot depend on choice sequences. Sequence of 'figures' (finite and proceeding infinitely) introduced in the process of constructing a spread M are called *elements* of M; spreads and their elements are dubbed *mathematical entities*.

Apart from these two means of generating of mathematical objects, as infinitely proceeding sequences or spreads, Brouwer acknowledges still another sort of mathematical objects, called *species*. As he puts it, a mathematical species is a property

(...) supposable for mathematical entities previously acquired, and satisfying the condition that if they hold for a certain mathematical entity, they also hold for all mathematical entities which have been defined to be equal to it, relations of equality having to be symmetric, reflexive and transitive; mathematical entities previously acquired for which the property holds are called elements of the species. (1952, p. 511)

Let us start in elucidating what here can be meant by 'property'. Property should not be thought of as given simply by announcing a predicate, since the concept of species is not meant as a means of introducing a mathematical object as the set of whatever things that satisfy a given predicate. That is, the ideology that underlies Cantor's comprehension axiom is rejected. The same holds true of Zermello's version of the axiom. (Brouwer 1919, p. 230)

The definition of species should provide a constructive analogue of the above notions, without implying that by announcing a property one thereby extends an inventory of mathematical objects. How then should we conceive of properties? Few places in which Brouwer tackles this question invariably refer to the possibility of embedding, or fitting-in, of a new system into an already constructed system. (1907, 1923) He mentions also a property as the possibility of rearranging a given mathematical system, that is, as changing an initial arrangement of elements of a constructed system. However, as the word 'supposable' suggests, we should not think of properties as already finished, or completed constructions of rearranging, or fitting-in, of one system in the other, but rather as hypotheses that such constructions are possible. If a system that characterizes a given property is embedded in an already constructed mathematical system, then the latter is called an element of the species.

We should view species as being introduced against the background of already constructed objects. In examining some object (or objects) O of this kind we find a feature of its (their) construction(s) and express it by saying that the construction(s) can be so-and-so rearranged, or some other construction may be embedded in it (them). Now we postulate the property P as given by the possibility of the above rearrangement or embedding of the chosen construction. Given that the construction of O can be so rearranged, or the chosen system can be embedded in it, we say that O is (or are) element(s) of P. For an object O to belong to a species P, it should be constructed and moreover, a proof should be known that an appropriate rearrangement of its construction, characteristic of this property, is possible. One further consequence is that the number of elements of a species may grow, as the property is considered against the background of an ever-richer stock of already acquired objects. It is perhaps better to say, as Brouwer did, "that during the development of intuitionistic mathematics some species may have to be considered as being tacitly defined again and again in the same way." (Brouwer 1981, p. 8) Finally, the appeal to 'mathematical entities previously acquired' is to guarantee that species is never an element of itself, which technically amounts to introducing a hierarchy of species. A species is of the first order if it is a property that only already acquired spreads and their elements can possess. A species of the second order is a property supposable for either spreads or their elements, or species of the first order. Higher-order species are defined inductively. As a result of these definitions, any species of the order n is also a species of the order $n+1$, but not the other way round. Finally, operations on species that are somewhat analogous to operations on classical sets are introduced, the result being a theory more subtle than the classical set-theory.[18]

We have introduced three forms a mathematical object can take in intuitionistic mathematics: infinitely proceeding sequence, spread, or species. It is perhaps worth remembering that the all-important concepts that gradually make it possible for the initially frugal land of intuitionistic mathematics to be inhabited by ever richer and more complex structures are the concept of infinitely proceeding sequence and its generalization, i.e. the concept of spread.

2.2. *What Is Not Intuitionistically Intuitive?*

Let us now turn to the 'negative' aspect of the intuition of two-ity, as it allows the intuitionist to dismiss some classically acceptable proofs and definitions on the grounds that they do not correspond to constructions that may be arrived at by following the intuition. To see what role the intuition of two-ity plays in Brouwer's rejection of some methods of classical mathematics, it is instructive to discuss his arguments. To this end, we will consider his objection to Cantor's definition of the second number class (in contemporary terms: set of ordinals of denumerable cardinality) and Cantor's famous diagonal proof. We shall start with the intuitionistic order relations, introduce ordinal numbers, and proceed to well-ordered species. This presentation is based on Brouwer's Cambridge lecture on order (1981, pp. 40–56).

I. Given that for elements of a species S relations of order $<$ and the identity $=$ are assumed, we call a species *simply ordered* if for any pair of elements x,y of which it is known that they are different, it can be decided that either $x < y$ or $y < x$. In a *completely ordered* species we require further that for any x,y it can be decided whether or not $x = y$. If two simply ordered species A and B can be brought into one-to-one correspondence f, such that for any x,y of A, if $x < y$ in A, then $f(x) < f(y)$ in B, these species are called similar or it is said that they have the same *ordinal number*. The ordinal number of a species that is similar to the species of natural numbers is designated by ω. Such species are called *fundamental sequences*.

II. To introduce intuitionistic *well-ordered* species we need first the notion of the sum of a species. Let R denote either a finite or a fundamental sequence of disjoint completely ordered species N_i, and M their union. We then say that

M is the *ordinal sum* of R if M is completely ordered and such that for any e', e'' from M the following obtains:

1. for e' and e'' belonging to different N_m and N_n, respectively, $e' < e''$ in M if $N_m < N_n$ in R;
2. for e' and e'' belonging to some N_k, $e' < e''$ in M if $e' < e''$ in N_k.

The operation of executing the ordinal sum M is called *addition*. Now, well-ordered species can be introduced inductively:

1. Any one-element species is well-ordered.
2. Given that in the stock of constructed well-ordered species a finite sequence of disjoint well-ordered species has been found, its addition

yields a well-ordered species (ordinal number). This addition of the finite sequence of well-ordered species is called the *first generating operation*.

3. Given that in the stock of constructed well-ordered species a fundamental sequence of disjoint well-ordered species has been found, its addition yields a well-ordered species (ordinal number). This addition of the fundamental sequence of well-ordered species is called the *second generating operation*.

Now, after becoming acquainted with the intuitionistic notion of the well-ordered set, we proceed to investigate Brouwer's reasons for rejecting Cantor's set theory, as he presented them in his dissertation (1907). As the creation of finite ordinals is unproblematic in the light of clauses (1) and (2), let us start with the introduction of ω. By intuitionistic standards the notion of the infinitely proceeding sequence or fundamental sequence is not controversial. So, given a fundamental sequence of natural numbers $1, 2, 3, \ldots$ the second generating operation yields a well-ordered species whose order type is denoted by ω. This species is added to the stock of available well-ordered species. By forming a two-element sequence (A, x) where x is a one-element species and A a species of order-type ω, disjoint from x, by executing the first generating operation one obtains a species of the order-type $\omega + 1$. Again, this species is added to the stock of well-ordered species. A further application of clause (2) to the sequence $(A, x_1, x_2, \ldots, x_k)$ where $A, x_1, x_2, \ldots, x_k$ are pairwise disjoint well-ordered species and every x_i is a one-element species, yields a well-ordered species of order-type $\omega + k$. Moreover, in the case of two well ordered disjoint species of order-type ω, the first generating operation leads to the construction of a well-ordered species of order-type 2ω. Further application of clause (2) yields the sequence of well-ordered species whose order-types increase as follows: $\omega, \omega + 1, \omega + 2, \ldots, 2\omega, 2\omega + 1, \ldots, m\omega + n, \ldots$ If these species are disjoint, their addition according to clause (3) gives rise to a species of order type ω^2. Proceeding further one obtains the fundamental sequence whose elements have the general form: $m_1 \omega^p + m_2 \omega^q + \ldots$, where m_i and $p, q, \ldots$ are natural numbers. Now, addition of this sequence of species results in a species of order-type ω^ω. A still further application of both methods of generation introduces well-ordered species of ever higher order-types.

One can notice that the above sketched method of introducing ever higher ordinals is not a point of contention between Brouwer and Cantor. Moreover, both of Brouwer's generating operations correspond to Cantor's two principles of creating ordinals, as explicitly stated in his 'Über unendliche lineare Punktmannigfaltigkeiten'.[19] Cantor's first creation principle permits the introduction of a new ordinal by means of adding a unit to an already given ordinal, while the second principle permits us to take as a new ordinal a limit of an infinite sequence of already defined ordinals. (*ibid.*, pp. 195–196)

Moreover, Cantor proves and Brouwer agrees that any order-type introduced by both the principles is denumerable and consequently, there are at any stage at most denumerably many ordinals. The point of contention is Cantor's unrestricted concept of totality that underlies his third *Erzeugungsprinzip*. This principle states that if all the already constructed ordinals satisfy a certain condition, then this condition can be imposed as a requirement to be satisfied by all ordinals to be built next; the totality of ordinals that fulfills this requirement is a new mathematical object.[20] The assumption of the third principle permits Cantor to define the second number class:

We thus define the second number class (II) as the totality of all numbers α that can be formed with the help of the two creation principles and that progress in the determinate succession $\omega, \omega+1, \ldots, v_0 \omega^\mu + v_1 \omega^{\mu-1} + \ldots + v_{\mu-1}\omega + v_\mu, \ldots, \omega^\omega, \ldots, \alpha, \ldots$, on which the requirement is imposed that all numbers preceding α, from 1 on, constitute a set of the cardinality of the first number class (I). (*ibid.*, p.197)

In the present terminology, we say that the above is a definition of the set of all denumerable ordinals. Brouwer, after quoting *in extenso* the above fragment, reacts to the proposed definition by claiming that the totality of all αs

(...) is something which cannot be thought of, i.e. which cannot be mathematically constructed (...). (1907, p. 81)

If we follow Brouwer's definition of well-ordered species, then at any stage the number of *all* already introduced ordinals is denumerable. On the other hand, it can be shown at any stage by constructing an appropriate ordinal that there are more ordinals than those included on the list of those already acquired. The word *all* as used by Cantor abstracts from this very mathematical fact that at no stage can both principles yield the totality of all numbers α. Thus, the totality is introduced by assuming a straightforward stipulation that all numbers α that satisfy a certain condition form a set. What, however, does Brouwer find wrong with this demand? Or, to ask a similar question, why does Brouwer reject the third *Erzeugungsprinzip*? According to him, for a totality to be introduced, it needs to be constructed by the rule 'and so on'; moreover, as he puts it, this rule should refer to an order type ω of equal objects. (*ibid.*) We may take this as a stance on where the rule is meaningful: it is meaningful as a means of generating of either a fundamental sequence or a finite sequence, on which Brouwer's two operations of generation of ordinals may act. Thus, the 'and-so-on' lies behind both of Brouwer's generating operations. We properly think of such a succession as represented by dots in an expression for a fundamental sequence of natural numbers, say, $1, 2, 3, 4, 5, \ldots$ Similarly, one may think of the *succession* as represented by the following expression:

$$\omega, \omega+1, \ldots, v_0 \omega^{mu} + v_1 \omega^{mu-1} + \ldots + v_{\mu-1}\omega + v_\mu, \ldots, \omega^\omega, \ldots, \alpha, \ldots$$

where again the dots refer to the intuitive 'and so on'. The mistake, according to Brouwer, lies in supposing that the dots occurring after α, while representing intuitive 'and-so-on', at the same time point to *all* denumerable ordinals greater than α. That is, on this misconstrual, one interprets the dots occurring after α and indicating denumerable ordinals greater than α as being on a par with the dots figuring in an expression standing for a fundamental sequence, say $1, 2, 3, \ldots$ Writing in this way, in other words, we pretend that the totality of all denumerable ordinals is given by a fundamental sequence, which of course it is not, as clearly shown by Cantor's proof that the second number class is not denumerable. Accordingly, 'the totality of all αs' cannot be thought of as generated by the addition of a fundamental sequence of already acquired well-ordered species. Someone who imagines the generation of the second number class to be on a par with generating a fundamental sequence misconstrues the rule 'and so on'. From Cantor's point of view, however, the above objections are hardly convincing. He will presumably agree on what one may see as the point concerning notation, namely that the dots hinting at all denumerable ordinals have another meaning than dots in an expression standing for a fundamental sequence. He will also agree that the totality of all denumerable ordinals cannot be arrived at by following the intuitive 'and-so-on' as his move of introducing the third creation principle, over and above the two others, clearly suggests. Thus, in this dispute over legitimacy of the second number class there are no easy winners. Each party has his own standards by which he introduces mathematical objects. The point of contention is Cantor's third *Erzeugunsgsprinzip*, as it allows the introduction of a mathematical object to whose legitimacy Brouwer cannot agree. The contentious principle is an application of Cantor's more general method, to which we soon turn.

It is easy to notice that the operation 'and so on' is closely related to the intuition of two-ity, so closely that it may pass for another name for this intuition. Thus, Brouwer's opposition to admitting the second number class as a legitimate object stems from the fact that that class cannot be constructed in accordance with the intuition of two-ity. As Brouwer sees it, the third creation principle that permits Cantor to introduce the second number class is an instance of a general method of forming mathematical objects, as stated in Cantor's 'Über unendliche lineare Punktmannigfaltigkeiten' and quoted *in extenso* in Brouwer's dissertation:

The method of correct formation of concepts is in my opinion always the same. One takes a thing without any property, which is at the beginning nothing else but a name or a sign *A* and gives to it in an ordered way many, maybe even infinitely many, different conceivable predicates, whose meaning in the light of already given ideas is known and which should not contradict each other. Thereby relations of *A* to other already given concepts are determined, namely to related concepts. If one is finished with this, all the conditions for waking up of the concept *A* that has slept in us are fulfilled and it steps ready-made into being (*Dasein*) with all that intrasubjective (*intrasubjektiven*) reality which can only be required of concepts. To determine its transient meaning is then a matter of metaphysics. (Cantor 1932, p. 207, tr. by A. Hüttemann)[21]

From Brouwer's standpoint, however, objects introduced according to the above method are of a logical nature only, they are not mathematical. This stance makes him reinterpret Cantor's diagonal proof that attempts to show that the second number class has higher cardinality to that of the totality of all natural numbers. On this construal Cantor's proof shows that both statements below are false:

1. The second number class is conceivable and denumerable. 2. The second number class is conceivable and there is a cardinal number between its power and that of the first class number. (1907, p. 82)

But in Brouwer's intuitionism neither the second number class nor its cardinality are conceivable mathematical entities. Given that these concepts do not lead to contradiction, Brouwer assumes they stand for 'logical objects'. They do not, however, have any mathematical meaning. Logic is for Brouwer, as we will see, a science concerned with regularities obtaining in linguistic systems. Thus we have a sharp distinction: on the one hand, mathematical objects, that is, constructions one can carry out by following the intuition of two-ity, and on the other objects introduced by logical or linguistic means only, which do not have mathematical status. However, in order to fully grasp the significance of Brouwer's claim concerning the logical but non-mathematical nature of objects that the classical mathematician works with, we need to study in more detail his ideas on the status of logic. We shall come to this issue in sections 3.4 and 3.5.

Had Brouwer not noticed that a source of paradoxes of Cantor's set theory lies in the unrestricted comprehension axiom, which is the heart of Cantor's method of introducing mathematical objects as described above, he would not have been a mathematical genius. One might thus expect him to amend the axiom, so that the paradoxes could not be deduced. However, one finds him rejecting altogether the whole approach, as based on a false assumption that mathematically meaningful theorems can be deduced by first introducing objects by means of predicates or some other 'linguistic' method, and then deducing from sentences concerning such objects with the help of rules of logic other sentences that allegedly have mathematical meaning. His diagnosis is far-reaching: not only is Cantor's method of introducing mathematical objects wrong; logical reasoning, if it does not correspond to constructible objects, is equally unreliable. Thus, while summing up his exposition of the Burali-Forti paradox he asserts:

(...) there is no reason to consider such a result as paradoxical: if we create logical structures without mathematical structures which they accompany as their linguistic counterpart, every such structure may be contradictory as well as consistent. (1907, p. 84)

2.3. *Some Objections to Brouwer's Concept of Intuition*

The idea of founding exact knowledge on a concept of intuition has traditionally evoked criticism. As we take it, the criticism objects to the assumption of a non-inferential faculty of knowing that allows a direct acquaintance with the intuited objects and to a belief, which sometimes goes together with this assumption, that this faculty is infallible. It seems, however, that since Brouwer's concept of intuition diverges considerably from its traditional counterparts, mathematical intuitionism escapes at least some sorts of the criticism leveled against intuitive knowing. To see that various sorts of criticism of founding our knowledge on intuition do not apply to Brouwer's epistemology, let us enumerate what for Brouwer the intuition of two-ity is and what it is not.

First and above all, the intuition is some means of introducing (or creating) mathematical objects. The mechanism of succession guarantees that to any mathematical object another one may be added. Following this intuition, one may create finite complexes of two, three, four, etc. elements. As the subject can think of such processes as proceeding without any end, and consider not only the outputs of processes, which are necessarily finite, but also processes themselves, he arrives at the concept of the infinitely proceeding sequence. Needless to say, such objects cannot be thought of as 'completed'. To ensure that the sequence proceeds without an end, the subject either has recourse to the concept of the law that determines any element of the sequence, or simply postulates that the process never stops.

Second, it is highly doubtful that the primary role of the intuition, as described by Brouwer, consists in the direct presentation of objects, that is the role the notion of *Anschauung* plays in Kant's philosophy. In a similar vein, the primary goal of Brouwerian intuition is not to provide something like direct acquaintance with mathematical facts. Also, it should not be compared with *seeing* in a direct and non-inferential way that a theorem holds. (Tieszen 1989, p. 2) The role of the intuition of two-ity is extremely limited: it only allows for mathematical constructions to be carried out. Thus, the intuition is like a mechanism which, if skillfully applied by the subject, allows him to arrive at any allowable mathematical construction. Of course, once a construction is carried out, it is also given to the mind, because it is a 'mental' construction. So, one may say, if anything is constructed by the intuition, it is 'directly given' to the mind. Thus, in Brouwer's system, there is no need for a mediating faculty, which presents constructed mathematical objects to the mind.

Third, the claim that it is 'intuition of time' needs a qualification. It is true that in Brouwer's philosophy time makes one sensation give way to the other. But, if the subject did not experience the passage of time, it would not be the mind and would not have the intuition of two-ity. If we may say so, the subject would remain in the state of pure consciousness, which does not differentiate between present and past sensations. Speaking precisely, however, we cannot even say that it is the subject, as this concept is understood in Brouwer's philosophy. Perhaps more importantly, time itself or rather its 'global structure' is a result of the causal attention and the intuition of two-ity. In this respect, it is no different from other physical objects. (Brouwer 1979, p. 401)

Fourth, one may concede that the faculty Brouwer introduces may be called 'sensible intuition', but only if this means that in every act of knowing that proceeds in the 'subject-object' schema, sensations must be presented as ordered by the relation of succession. But thoughts and other constructions of the subject must equally be so arranged. That is, anything the mind experiences in this schema is arranged according to one form: the form of succession, or two-ity, as Brouwer puts it.

Fifth, although one may introduce a number of other intuitions as well, say those allowing the subject to separate a mathematical entity, or mentally differentiate between objects, Brouwer does not identify the intuition of two-ity with any of them. Even more, he does not even mention any such faculty in his writings. Whether these faculties are needed in his doctrine is separate question.

Finally, only indirectly is the intuition of two-ity a means of acquiring a mathematical insight. It would be very imprecise to say that one intuits a mathematical object, say a triangle or a number, and explain this notion traditionally, that is, as complete, direct and simple knowledge of the object. Obviously, the intuition plays an epistemic role, as it carries out the appropriate constructions of mathematical objects, but it would be incorrect to assume that knowledge by intuition has such distinctive features. In Brouwer's framework, the intuition does not even ensure that the acquired knowledge is certain and error-free. In mathematical practice, faculties other than the intuition play a role as well, with the memory being especially important. And, for Brouwer, it is precisely the memory, with its finitude, that is responsible for mistakes. (1933, p. 443) So, intuitionistic mathematical constructions are subject to mistakes and confusion, as even the simplest operations on them involve their retention in memory.

Similar reasons indicate that Brouwer's concept of intuition can hardly be compared to an inner voice that helps you to guess correct answers to mathematical puzzles, like the one Wittgenstein occasionally attacks in his *Philosophical Investigations*. Intuition is identified there (§213) with a faculty of guessing and choosing a law of the progression of a sequence of numbers given that its initial segment is known. That is, given such an initial fragment one can, in principle, interpret it in a number of ways, for example by finding various arithmetical expressions that describe the whole infinitely proceeding sequence. One may perhaps complain that Brouwer should have assumed an ability that accounts for the performance of such tasks. But why should this be identified with what he called 'intuition'?

Apart from the objections that assume that Brouwerian intuition is a faculty of knowing, in a direct, non-inferential way, and which can be dispelled as based on misunderstanding, there are others more directly linked to the present project. It may be argued that founding mathematics on the intuition of two-ity leads to the question as to how there can be agreement between mathematical results obtained by different mathematicians, each of them following his 'own' intuition. A similar objection, but applied to counting, relies on asking whether intuition can always bring forth a new unit. To reply to these objections, one needs first to note an intimate link between two basic concepts of Brouwer's philosophy: mind and intuition. The connection is as follows: the mind experiences an ordered flow of sensations, in which any sensation is followed by another. Accordingly, if sensations are not ordered, then what receives them is not the mind. (This may be the consciousness, though.) On the other hand, if you focus on a phenomenon of one sensation following another and strip off their sensuous content, you obtain the empty form of the succession which is called 'the intuition of two-ity'. Hence, the intuition of two-ity is almost a defining feature of the Brouwerian mind: if something is the mind, then it has the faculty called 'the intuition of two-ity'. On the other hand, there can obviously be no intuition without the mind whose ability it is. Such an intimate link between the mind and the intuition of two-ity makes Brouwer's epistemology almost immune to criticism based on empirical data. Even if psychology were to discover a human being who does not experience an ordered flow of sensations and whose mathematical intuition sometimes stops bringing forth new units, this would not undermine Brouwer's philosophical views. The Brouwerian intuitionist would likely deny that such a human being has a mind, in the way this notion is understood by Brouwer. In the same vein, somebody who does not perceive a flow of sensations (because he is asleep or has taken drugs) cannot be said to have the Brouwerian mind. Moreover, in this framework it is difficult to imagine two mathematicians obtaining different answers to the same mathematical problem, since if the difference is not attributable to a flaw of a memory, it must result from a 'gap' in the functioning of the intuition of two-ity. However, such a 'gap' can only mean that at least one of the mathematicians has ceased to be (or have) the mind as Brouwer defines it, which has in turn brought about the discrepancy in their results. Thus, the conclusion is, since the intuition is a defining characteristic of the mind, that any mathematician is capable of repeating the constructions of the other as long as he has the mind.

Such a reply, following the fundamental concepts of Brouwer's philosophy, is nevertheless very speculative and ambiguous. One may wonder, for instance, whether Brouwer's concepts of mind and knowing subject should be interpreted along Kantian lines, as a transcendental ego, or rather in a more subjectivist fashion. But whatever interpretation one chooses, the concept of mind will necessarily diverge from that of the group of faculties of a flesh-and-blood mathematician, faculties that are responsible for that person's mathematical reasoning. Thus, even if one might be satisfied with this or that interpretation of Brouwer's concept of mind, there would still be a lingering question; namely, what is the relation of the psychic subject to the epistemic subject? How are the abilities of a real mathematician linked to the mind that

44   CHAPTER TWO

Brouwer talks about? These somewhat airy questions appear to be related to another query, namely how one should understand Brouwer's notion of construction.

2.4. *The* Possible *Construction: Brouwer's Notion of Possibility*

As has already been mentioned, a statement holds for Brouwer only if its proof *is*, or *can* be, carried out. This formulation leaves open questions of whether the 'can' is really needed and if so, how it should be understood. On the other hand, as constructions are said to be effected by a subject, we need a clarification of what this subject is. Is it a flesh-and-blood mathematician, an idealized mathematical community or, even more abstractly, an ideal transcendental ego? Both issues, i.e., interpreting the intuitionistic possibility and getting a grip on what the subject is, are to some extent related. For instance, if the subject is assumed to posses the ability to decide the truth or falsity of any mathematical statement involving quantifiers by sheer introspection of its instances, which may be infinite in number, the rationale for claiming that he *can* neither prove nor disprove a statement simply disappears. Accordingly, one may hope that by explaining Brouwer's concept of possibility one will get an insight into how to understand the intuitionistic subject. To proceed, let us limit our attention to statements that have mathematical meaning, that is, statements that are understood as claiming the existence of the appropriate intuitionistic constructions accessible by following the intuition of two-ity. Secondly, we should rather restate Brouwer's position in terms of assertability than of truth, although he used the notion of truth. By doing so, we avoid discussing the intricacies of Brouwer's concept of truth at this point, a subject to which we return later. We also assume here without discussion the intuitionistic interpretation of disjunction according to which a disjunction is assertible only if at least one of its disjuncts is assertible. Bearing these restrictions in mind let us consider two examples, of which the first elucidates intuitionistic reservations to non-constructive definitions, while the other relies on a mistaken view of Brouwer's mathematics.

H The number k is defined as follows: k is equal to the largest prime number l, such that $l-2$ is also a prime number. If there is no such a number, let $k = 1$. (Heyting 1956, p. 4)

B The number 1000^{1000} is equal to a sum of two prime numbers, or it is not equal to a sum of two primes.[22]

The difference between the two examples is as follows: it is not known whether there is the largest prime number l such that $l-2$ is also a prime; this number is sometimes called the largest twin number. Put in intuitionistic terms, neither the method of calculating the largest twin number nor a proof that derives a contradiction from the supposition of carrying out its construction is known. In the case of **B**, on the contrary, although the problem has probably not, as a matter of fact, been solved by anybody, the procedure for its solution is nevertheless known. Now, intuitionists explain that in order for the condition **H** to succeed in defining the number k, either that number should be calculated, or at least an effective method of its calculation should be known. As none of the above is known, they claim that definition **H** cannot be accepted. On the other hand, in example **B** the disjunction can be asserted on the grounds of the knowledge of a procedure that shows for any number whether or not it is a sum of two primes. This shows that by intuitionistic standards the requirement of actually carrying out of a construction is too strong a demand for the assertability of the respective sentence. Moreover, this demand has odd consequences since, in the case of some sentences, the question of whether or not a sentence is assertible would depend crucially on the subject's decision either to perform rather trivial, though cumbersome calculations, or not. A report on what sentences are assertible would be a factual record of someone's decisions about which mathematical problems are worth solving. Accordingly, assertability would not be so much a property of sentences, but rather something that characterizes both the sentences and an actual person (or a community).

The moral of this story is that in characterizing intuitionistic assertability one needs to take seriously the little word 'can'. What then does it mean that a construction can be carried out? On the one extreme, one may say that as long as any mathematical problem is restated in intuitionistic terms, its solution is 'determinate', and support this with a claim that the intuition of two-ity is so rigid that it does not leave room for genuine mathematical statements that *can* never be either proved or disproved. With this stance the rationale for anti-realism seems to be gone: for instance, its proponent does not have any reason to object to the correctness of definition **H**. Thus, if we want to stay with the intuitionist we must rebuke the above suggestion. This leads to a view that the intuitionistic 'can' should be interpreted epistemically: it refers to available mathematical knowledge, and not to the ontology of the hypothetical totality of all conceivable mathematical constructions. However, this formulation offers only a slight improvement, as it shifts the problem into explaining what 'available' means in this context. Nevertheless, with this proposal the phrase 'a construction can be carried out' should relate to abilities of a subject that has both the intuition of two-ity and the appropriate mathematical knowledge. Such a characterization makes irrelevant any factual obstacle that may hinder the execution of proofs. For instance, although it is conceivable that an imminent cosmic catastrophe would bring the world to an end before any computer executes a program designed to show whether or not 1000^{1000} is a sum of two primes, the intuitionist does not object to the assertion of the sentence:

1000^{1000} is or is not a sum of two primes.

Physical events have no bearing on what can be asserted in mathematics. However, the proposal leaves a lingering question as to how to characterize the mathematical knowledge relative to which some constructions can be effected and others not. Suppose that Fermat, as a matter of fact, had proved his Last Theorem and moreover that his proof met intuitionistic standards. If this were so, Fermat could assert the sentence:

For all natural numbers x, y, z and $n > 2$, $x^n + y_n \neq z^n$.

On the ground of his proof he could as well assert an instance of the principle of the excluded middle to the effect that the theorem either holds or not. However, with the knowledge of Fermat's proof lost, the intuitionist could assert neither the sentence stating the theorem nor the above instance of the principle of the excluded middle. Yet, on the supposition that the Andrew Wiles' recent proof of the theorem is constructive, both the theorem and the discussed instance of the principle of the excluded middle are again assertible by intuitionistic standards. Other hypothetical stories can be told as well; they will introduce amateur but unknown mathematicians or mathematical communities in other cultures that have known or know proofs or disproofs of statements which at present are believed to be undecided. All such stories point to a certain oddity, namely that a sentence may become assertible, but may also cease to be so. They suggest that any mathematical problem whose solution is not known at present *could* have been solved in the distant past or its solution may be known even at present by members of a mysterious though remote mathematical community. Taking this optimism seriously, if one concedes in an explication of what mathematical knowledge is that it includes any mathematical discovery ever made, and which may be known or not at present, then much of the motivation for objecting to the assertability of instances of the excluded middle is gone. To that, however, one may object that because of truly undecidable mathematical problems, the belief that any given mathematical statement could have been decided is far too optimistic. Thus, at this point, the debate hinges upon the existence of truly undecidable mathematical problems. We may conceive of a position that partially agrees with that of the intuitionist on his construal of the principle of the excluded middle, namely, as the claim that a mathematical problem can either be proved or disproved, but objects to the intuitionistic notion of possibility. This position would accuse the intuitionist of operating with too narrow a concept of possibility, which is related either to a particular mathematician, or to a particular community of mathematicians, and would optimistically claim, as Hilbert once did, that any mathematical problem is solvable.[23] As a result of such a combination of views, the resistance to the principle of the excluded middle disappears. The intuitionist can take at this point one of two courses of action. First, he may try to defend his notion of possibility by showing that all the stories about ingenious but unknown mathematicians are irrelevant to his concept of assertability. Secondly, he may scoff at belief in the solvability of every mathematical problem as groundless optimism. At first sight the latter option seems cogent, since if we shift our attention from an intuitive notion of solvability to the formal concept of decidability, the existence of truly undecidable problems is a straightforward consequence of Gödel's incompleteness theorem. However, formal decidability is a concept relative to a given formalization of a mathematical theory, and consequently, the fact that some sentence is undecidable in a formal theory does not give any hint as to whether it is intuitively solvable. The question of the solvability of any mathematical problem cannot therefore be answered along these lines. Moreover, the debate between the intuitionist and his opponent depends crucially on a decision as to whom the burden of proof rests with. Should the intuitionist's opponent back his optimism concerning the solvability of mathematical problems by some argumentation, or rather does the *onus probandi* lie with the intuitionist, that is, should he demonstrate that there are truly undecidable problems? There is a natural inclination to say that the burden of proof rests on the intuitionist's opponent, since his epistemic optimism is so extreme that we need some testimony of it. If so, the prospects for intuitionism look better, since no such argumentation is known and it is difficult to conceive of one. However, the intuitionist's position is rather delicate in this respect. He cannot argue that there are truly undecidable problems and, moreover, it is problematic whether he can ascribe any meaning to this notion. In his framework, a problem is truly undecidable if it can never, on mathematical grounds alone, be either proved or disproved. But to assert that on mathematical grounds alone the statement p can never be proved is tantamount to asserting its negation, that is not-p, since the former assertion means that any proof of p is impossible. In a similar vein, the assertion that p can never be disproved boils down to asserting its double negation, i.e., not-not-p. Thus, on the intuitionistic account, the assertion that a statement is truly undecidable, if at all meaningful, amounts to uttering a contradiction.[24] Accordingly, the intuitionist's moves in the envisioned debate must be rather prudent. On the one hand, he needs to maintain skepticism as to the solvability of any mathematical problem and require from his opponent evidence for his epistemic optimism. On the other hand, he should abstain, on pain of making himself incoherent, from announcing that some problem is truly undecidable. While arguing that an instance of the principle of the excluded middle is not assertible, he will maintain that it is not decided, or better, that it *cannot* be decided in the present state of knowledge whether p or not-p holds. And, when asked whose state of knowledge this is, or how he knows that the problem under consideration is not decided in such a state of knowledge, he will have recourse to some arbitrary idealization. On the one hand, the intuitionist may say that this knowledge consists of all mathematical constructions known at present to some community of mathematicians. On the other hand, he may invoke an ideal subject who is capable of carrying out any known construction. Both solutions are troublesome, as they make assertability depend on contingent factors of the historical development of constructivist mathematics. But these idealizations are still worth assuming, since otherwise we would have an entirely subjectivist concept of assertability that makes this notion dependent on the mathematical abilities of a flesh-and-blood mathematician.

To sum up this discussion briefly, in characterizing assertability the demand of actually carrying out a construction is too strong and leads to odd consequences. Thus, we are left with a not too satisfactory explanation that appeals to *possible* construction, where this possibility is epistemic, related to the mathematical knowledge available at present and independent of physical events.

3. LANGUAGE, TRUTH AND RELATIONS BETWEEN LOGIC AND MATHEMATICS

3.1. *Language*

Brouwer took an interest in language for at least two reasons. The first was rather negative but strictly linked to his mathematical program, because one of the most salient features of his intuitionism was a distinction between, on the one hand, mathematics viewed as a multitude of mental constructions, and, on the other, the linguistic expression of those constructions. Accordingly, he aimed at clearing mathematics of merely linguistic means of introducing mathematical objects. Rejection of such methods lies at the heart of Brouwer's opposition to Cantor's set theory, Zermello's set theory and the formalism of David Hilbert that was emerging at that time. Brouwer opposed Hilbert's project since he believed that studies of systems of signs have little in common with the 'real' subject of mathematics. The same view on the role of language in mathematics motivated his rebuke of logic, because, as he saw it, logic deals only with linguistic regularities obtaining between assertions.

On the other hand, Brouwer was one of the founders of and later an active participant in the movement known as 'Significs', whose aim may be roughly described as studies of various ways in which linguistic expressions act on people. As he recollects in 'Synopsis of the signific movement in the Netherlands' (1946), the history of the movement harks back to 1915, when in the thick of wartime propaganda a group of Dutch intellectuals came to the conclusion that the future peaceful coexistence and collaboration of nations required setting up an institution for the thorough investigation of languages. Witnessing the use of militant slogans and propaganda attacks, they realized that European languages contain words or expressions that can easily be used to propagate hatred and aggression. This observation led to a project for investigating the ways in which linguistic expressions invoke people's emotions; one reason for such studies was to uncover words that promote hatred. The group also dreamed that it would be possible

[T]o coin words of spiritual values for the languages of western nations and thus make those spiritual values enter into their mutual understanding(...) (*ibid.*, p. 465)

A part of this program was to tag expressions that under the disguise of appealing to some higher values bring about aggressive desires and the will to seize other people's property and to enslave other human beings. Similarly, words falsely suggesting spiritual values, where in fact only material safety and comfort are concerned, were to be identified. The linguistic reform was to be accompanied by legal reform, as the group intended to track and combat those elements of the legal system which lead to a regress of spiritual values. They also recommended "appropriate limitations for the sphere of influence of law and technics." (*ibid*, p. 465) The preparatory manifesto was signed in 1918 by four people, including Brouwer. Soon, others joined the founders; the group then consisted of F. van Eeden, H. Borel, H.P.J. Bloemers, G. Mannoury, L.S. Ornstein, J.I. de Haen, and Brouwer.

Brouwer's 'Synopsis of the signific movement in the Netherlands' makes it clear that the leaders differed in the emphasis they put on various parts of the program, which was the source of the eventual decay of the movement. Among their various evaluations of the aims of significs, Brouwer's standpoint distinguished itself as the most ambitious. He still held his initial conviction that significs should aim at the reform of European languages by eliminating harmful vocabulary and coining words appropriate for the expression of spiritual values. Nevertheless, his enthusiasm for significs faded with the passage of time.[25]

Although Brouwer's signific papers show his eagerness for a reform of European languages, they tell little about his views on language and linguistic communication. Both these subjects, however, were discussed in his most general papers of 1929 and 1948, which cover a variety of topics from metaphysics to examples intended to show the invalidity of classical logic. Such a placement is significant, as it hints at the role his conception of language was intended to play in the proposed revision of mathematics. Brouwer's remarks on the nature of language start in both papers with an account of the origin of language. As in the case of the emergence of the objective world, the emergence of language is rendered possible by a faculty called the *causal attention*. To emphasize that this faculty is directed at linguistic phenomena (broadly understood), he uses the term *linguistic causal attention*. (1948, p. 482) Since the activities of this faculty consist in postulating mathematical structures in which sense data are embedded, he also uses another name, that is *mathematical attention*.[26] The key concept of Brouwer's account of linguistic phenomena is *transmission of will*. Transmission of will takes place even in the most primitive stages of civilization, whenever some individuals coerce others into performing their jobs and whenever humans cooperate in fulfilling various tasks. Vehicles of will-transmission at such early stages are simple gestures, grunts and unarticulated sounds. However, as jobs and tasks become increasingly complicated and the organization of human groups more complex, these simple means of communication stop being satisfactory. At this stage the totality of objects, deeds, laws, actions and events that are involved in the process of communication are subjected to the linguistic causal attention. This results in the emergence of a mathematical system, whose basic elements are denoted by *basic linguistic signs*. On the other hand, the mathematical system imposed on the totality of communication phenomena contains certain rules, which govern what relations obtain between its basic elements, these being supposedly identified with grammatical rules. The grammatical rules operating on basic linguistic signs allow one to articulate and transmit the most complicated commands and wishes. People receive a splendid vehicle of will–transmission, suitable to the expression of a huge variety of will. (1929, 1933, 1948)

From this rather fictitious account, Brouwer draws an immediate consequence by saying that language is a result of the social activity of humans and thus necessarily has a social character. (1929, p. 421) He goes even further and adds that if a person isolated from his or her community uses language as a means of supporting his memory, he takes into account knowledge of the community as well as its organization in doing so. Although Brouwer's remarks about linguistic communication are too sketchy to risk any definite interpretation, the impression is that two decades in advance of the publication of Wittgenstein's 'private language argument', the Dutch mathematician already held views directly opposing both the conception of language defended in *Tractatus* and the 'mentalist' picture of language as primarily a vehicle of thought and a means of supporting memory, of which the *locus classicum* is Locke's *An Essay Concerning Human Understanding*.

What, however, is the relation between the above story of the origin of language and Brouwer's philosophy of mathematics? To start with, it is necessary to note that for Brouwer the language of mathematics serves the same aim as other languages. Namely, it is part of an elaborate net of will-transmission. It is nevertheless somewhat troublesome to assume that in a discourse among mathematicians discussing, say, a proof of a theorem, one person literally transmits his will to another, presumably with the help of uttered sounds and gestures. What Brouwer likely has in mind in subsuming the language of mathematics under this broad category, is that the sole role of this language is to guide the hearer into an understanding of the thoughts of a speaker. He seems to suggest that the language does not have the other roles traditionally ascribed to it, such as for instance being an adequate formal representation of mathematical reasoning. Perhaps his picture of language can be made more familiar if one realizes that we sometimes come across an excellent math student, who is nevertheless very incapable of elegant, not to mention formal, expression of his thinking. That person may still be surprisingly successful in coveying his ideas by means of sketchy drawings, gestures, or metaphors. In such a case, one may be inclined to say that there is nothing peculiar to the language that is used to discuss mathematical matters. It is neither more precise nor more loaded with logical devices than other parts of our ordinary language. On Brouwer's account mathematicians follow the intuition of two-ity to create various mental mathematical systems. In such mathematical activities, a mathematician may or may not have recourse to some linguistic means. All told, however, language plays only a subsidiary role, as a way of supporting the memory. Then as soon as people attempt to communicate their mathematical results, the language of logical reasoning comes into being. But still the communication consists in invoking in other people "copies of mathematical constructions and reasonings" with the help of sounds or written symbols. (1907, p. 73) In this way some elements of mathematical systems are given their names, though the question of which parts of mathematical systems words correspond to receives various answers depending on the milieu of the mathematicians. (*ibid.*) Consequently, the language of mathematics is viewed as a rather fallible means of mathematical communication, sharply separated from mathematical constructions themselves.

3.2. *Against Hilbert's Program*

For anybody holding such ideas, the program of the formalization of mathematics must appear mistaken or even absurd, since formalization is then viewed as primarily concerned with studies of symbols occurring in reports on the mathematical activities of a subject or subjects. According to this interpretation, a formalist attempts to ascertain some theorems about mathematics itself on the basis of considerations pertaining only to a linguistic counterpart of mathematics. From this perspective, Hilbert's Program appears to consist in building, on the base of intuitive mathematics, systems of signs whose rules and operations are subject to mathematical investigations, which again are expressible by another system of signs, and that system is again studied mathematically. Accordingly, Brouwer opposed not only Hilbert's concentration on the language of mathematics, instead of mathematics itself, but also argued that his results concern only a system of symbols of a relatively high order. To present Brouwer's opinion on Hilbert's Program, let us recollect the central tenets of the project. Its core was a proposed method of justifying mathematics by finitary means. The justification aimed at saving the integrity of the whole of classical mathematics together with the concept of actual infinity, which was viewed by many as a source of paradoxes. Accordingly, Hilbert believed that what requires justification are proofs and theorems involving actual infinity, whereas the finitistic part of mathematics was taken to be unproblematic. The object of the studies was a system containing classical logic, set theory, arithmetic, and analysis. The system was supposed to be formalized, so a list of symbols together with syntactic rules determined which formula is well-formed. Also, some formulae of the formal system were selected as axioms. This together with syntactic rules of derivation made it possible to define, in purely syntactic terms, the concept of proof as an appropriate sequence of strings of signs. The intuitive division into problematic and unproblematic mathematics was expressed by a distinction between ideal and real sentences. The latter are meaningful, as they refer to concrete objects and accordingly do not stand in need of justification. The ideal sentences, on the contrary, invoke infinities, which makes their meaning spurious. As to accepted axioms and rules of inferences, it was assumed that they suffice to solve any mathematical problem that is expressed by a real sentence of the formalized language. Hilbert believed that any proof of any real sentence that proceeds *via* ideal sentences (or, what boils down to the same thing, appeals to ideal objects) can be transformed into a proof consisting only of real sentences, that is using only finitistic methods. In this sense, the ideal sentences or infinitistic part of mathematics would play only an auxiliary role. The argument for the possibility of such transformation of proofs, or as we say, for the *conservativeness* of mathematics, depends on the possibility of a purely syntactic characterization of proofs: these are treated as sequences of strings of symbols. Apart from the objective of proving the conservativeness, the other aim was a proof of the consistency of the system, that is, a proof that there cannot be two formal proofs in the system such that one ends with a formula, while the other ends with its negation. In order to prove consistency and conservatism, Hilbert introduced a new discipline that consists in investigations of formal proofs. The crucial requirement of the program was that conservativeness and consistency should be proved by finitary means only.[27]

Let us now look at Hilbert's Program from Brouwer's perspective. Hilbert starts with intuitive mathematics, or more precisely, with arithmetic, analysis, and some elements of logic. In Brouwer's terms, this first stage is the 'pure construction of an intuitive mathematical system' which, if it is applied to the external world, is a form in which external data are organized. (1907, pp. 94–95) Because we usually speak or write about mathematical systems, we have as the second stage a linguistic counterpart of mathematics that consists of sounds or written symbols. At the third stage one takes a mathematical view of the system of symbols and notices that the behavior of linguistic expressions is rather regular, as they are built according to some principles and are operated in accord with some rules. So, this stage can be seen as the beginning of a process of formalization of the language: symbols are divided into categories, there are rules that state what expressions are well-formed and what derivations are correct. The important thing is that at this stage linguistic expressions are still assumed to relate to mathematical structures. It is in the next, fourth phase that the meaning of the above linguistic structures and symbols is neglected while on the other hand the regular behavior of linguistic expressions is imitated by relations obtaining between elements of a newly constructed mathematical system of the second order. At this stage, mathematical reasoning need not be accompanied by a language which accounts for these elements. Since the system is liable to contradictions, the main task is to secure its consistency. According to Brouwer, the mathematics of the fourth stage is what proponents of the reduction of mathematics to logic are concerned with when they reason mathematically about the logical rules and principles distilled from the language that accounts for intuitive mathematics. Presumably, manipulations with Hilbert's formal proofs belong to this stage as well. One enters the next, that is the fifth stage once the constructions of the system of the second order are being explained in the accompanying language. An example of this is Peano's account, in plain language, of mathematical reasoning carried out in his formal system. In the next, sixth phase, this accompanying language is in turn subjected to mathematical studies. According to Brouwer, this takes place in Hilbert's program, in investigations of how formal proofs, viewed as linguistic structures, develop in accord with logical and arithmetical principles. Although the meaning of most expressions is again being ignored, (that is, what structures of the mathematical system of the second order the expressions refer to is not taken into account), some words like *more, two, progression* relate to the appropriate elements of the mathematical system of the second order. This disappears in the next, seventh stage, in which the meaning of linguistic structures is neglected and their relations are imitated by relations obtaining between elements of a newly constructed mathematical system of the third order. This stage corresponds to mathematical investigations of formal proofs, whose objectives are proofs of conservativeness and consistency. In the final step, these mathematical investigations are expressed in a language giving rise to constructions carried out in the system of the third order. These constructions are then interpreted as proofs of consistency or conservativeness. (Brouwer 1907, pp. 94–95)

The mechanism of introducing the hierarchy of mathematical systems relies on two 'transitions'. In the first, one turns from mathematics to a language in which it is described or accounted for, while in the second, after the language is formalized and meaning of its expressions forgotten, the formal structure that emerged is itself put under mathematical scrutiny. Accordingly, we have intuitive mathematics, then the mathematics one deals with while carrying out formal proofs, and finally mathematical reasoning in investigating formal proofs, that is, reasoning of proof theory. In Brouwer's view, however, both mathematical systems, of second and third order, relate strictly speaking to regularities of linguistic expressions and procedures. Thus, he sees a principal drawback of Hilbert's Program in its consisting in studies of linguistic structures that accompany mathematical constructions, and not of mathematics itself. And as he argues, the stages from the third on are deprived of mathematical significance, since mathematics itself is considered only in the first phase and somewhat accidentally in the second. Moreover, the results which obtain in the later phases concern, strictly speaking, only the mathematical systems of higher order. Thus, he claims, it is doubtful whether they have any significance for intuitive mathematics, as it occurs in the first stage.

Brouwer's conclusion may strike one as far too cautious. His enumeration of the stages inherent in Hilbert's Program seems to be rather noncontroversial, and probably only the most thoroughgoing formalists would object to it. Others would likely see in it a quasi-psychological account of what happens in the process of executing the program. Thus, a disturbing problem for someone who agrees with this account is to understand Brouwer's reasons for holding that a proof carried out in the third order mathematical system does not give information about intuitive mathematics. After all, given that we had a proof in a third-order system that we interpreted ordinarily as showing conservatism, this result, we are inclined to say, would concern formal proofs (which belong to the second-order system)—that is, formalized mathematics. Moreover, if someone is convinced that a given formalization correctly captures an intuitive mathematical theory, he may see in this result information concerning this (intuitive) theory. For instance, a proof of conservativeness may assure such a person that an appeal to actual infinities in that intuitive theory does not give rise to paradoxes. Brouwer, however, insists that the result pertains only to a language in which manipulations with formal proofs are described, a language which is deprived of mathematical meaning. He maintains, consequently, that the results pertaining to language do not give any hint about mathematics itself.

3.3. *Brouwer's and the Axiomatic-Deductive Method*

At this point a suspicion arises that it is not merely Hilbert's Program, but perhaps the axiomatic-deductive method, or still more generally, the turn from intuitive mathematics to its characterization in language, that Brouwer objects to. Thus, we need a clarification as to what tradition in the development of mathematics Brouwer was inveighing against and also of how deep this tradition runs. What emerges as the principal issue in his discussion of Hilbert's project is the relation between intuitive mathematics and language. If we focus on language, we easily discern its first function in regard to mathematics, namely that it is a means of making other people understand mathematical concepts, arguments or proofs. At this point it should not concern us whether Brouwer is right in advocating a disorderly picture that makes communication in mathematics close to invoking mathematical concepts in other person's mind by a process comparable to jostling or pushing. For what one usually has in mind while hearing the expression 'language of mathematics' is not a fragment of ordinary language in which mathematicians communicate, but such creations as the systems of Euclid's *Elements* or Peano arithmetic, or perhaps some contemporary formalized and axiomatized mathematical theories. Clearly, these are not very good as means of actual communication: you can hardly teach a child arithmetic by acquainting him with Peano's axioms. Thus, there is an inclination to say that the role of language that Brouwer puts so much emphasis on, namely as a means of communication, is neither the only nor the central one. What then are the aspects of the language of mathematics that Brouwer ignored? The first aspect consists in adequately *representing* mathematical concepts of intuitive mathematics and expressing what are conceived of as propositions of intuitive mathematics or someone's thoughts about mathematical matters. Secondly, and more importantly, it is assumed that the language has the resources required to represent our intuitive mathematical reasoning formally. That is, intuitive reasoning is first to be expressed in language, then its expression is to be elaborated so as to display a certain linguistic structure, and finally, if this structure is an instance of some scheme of valid arguments, the reasoning is to be judged as correct. Clearly, this latter task is more far-reaching than the former, as not only does it deal with expressing our thoughts by linguistic means, but it is also an attempt to prescribe, in terms of *forms* of linguistic expressions, which intuitive reasonings are correct and which are not.

Now, if we look at these two aspects somewhat ahistorically, we may say that they are merely postulates of a highly philosophical program, since after all in many of our disorderly discourses about mathematical matters it is hard to discern both precision in the expression of thoughts and devices that could be used to represent our intuitive arguments formally. Thus, an ideal language of thought and reasoning does not lie, so to speak, on the surface of our practices of linguistic communication and it is rather a postulate that our language, as used in communication, has at its deeper level the desired structure adequate for representing thoughts and prescribing reasoning. Alternatively, instead of postulating that this language may be extracted from mathematical discourses, one may set the task of designing a language which perfectly matches our thoughts and prescribes what reasoning is correct. The ideal of such a language is well situated in a grand tradition, intimately connected with such names as Plato, Leibniz, and Frege, that put a strong emphasis on language seen as a necessary condition of abstract thought.

If these postulates were only a philosopher's ideas, one could object that they are unrealistic or mythical. In such a case, displaying the disorderly way in which people sometimes communicate about mathematics would undermine the philosopher's postulate of an ideal language of thought and reasoning. However, both these postulates, of the adequate representation of our mathematical thoughts and the linguistic characterization (and, perhaps, prescription) of our intuitive reasoning, are so intimately linked with the actual history of mathematics that it is hard to conceive of a purely intuitive mathematics whose propositions are expressed without invoking a symbolism that attempts to represent their structures. Similarly, it is hard to imagine a mathematical theory whose reasonings are not expressed in such a language and validated on the grounds of their being instances of some schemata of valid arguments. What may come to mind as possible candidates for such purely intuitive theories are perhaps a pre-Euclidean geometry, or some other mathematical theory in an early stage of its development or indeed Brouwer's peculiar way of doing mathematics. Brouwer's mathematical practice reflected, after all, his negative assessment of formal and logical methods in mathematics. Despite the fact that the introduction of a symbolic notation would often have made his lectures or publications more concise, he preferred an informal style, relying on ordinary language and somewhat rarely having recourse to symbolism. In the few photographs taken at his lectures one sees him in front of a blackboard covered only by sketchy drawings, without the slightest trace of formulae that could remind us of axioms or definitions of newly introduced notions. Brouwer unquestionably took seriously his own vision of invoking mathematical ideas in other people's minds. He lectured with the help of informal means like drawing and preferred ordinary language, used with elegance and exactness. His way of teaching likely harked back to his disbelief that a more formal presentation could better facilitate his audience's grasp of the subject. Nevertheless, it is debatable whether this way of practicing mathematics was derived from a philosophical position he might hold, namely, that any presentation of a mathematical theory as an axiomatic system is inadequate, or was simply a result of his personal preferences.

To return to the concept of language as a medium of reasoning, it crucially includes rules of reasoning understood as regulating what sentences can be asserted on the grounds of acceptance of some other sentences. It needs to be assumed that sentences express 'moves' of intuitive reasoning. Now, in order for the rules to be applicable, their application cannot draw on a feature peculiar to any particular sentence; we need to have means of classifying sentences in terms of their structures so that the rules draw only on the structure or form of sentences. This is commonly handled by distinguishing a few expressions (operators), in terms of which the form of a sentence is explicated by stating what the principal operator and the component(s) of a given sentence are. These expressions are elevated to the status of logical constants; our initial rules of reasoning (or some of them) turn into familiar rules and laws governing logical constants. Now, the important point about any workable language of reasoning is that it must contain some logical constants or other, as well as rules and laws that regulate them. Accordingly, in order to see any attraction in this linguistic approach to reasoning one should conceive of logical notions, like those of conjunction, implication, negation, or contradiction, as having meanings independent of the sort of discourse. Have they, however, any meaning for Brouwer? In a famous passage (1907, p. 73), while addressing the logically-minded, he distances himself from anyone who thinks of mathematics in terms of contradiction, implication, and other logical notions:

The words of your mathematical demonstration merely accompany a mathematical *construction* that is effected without words. At the point where you enounce the contradiction, I simply perceive that the construction no longer *goes*, that the required structure cannot be imbedded in the given basic structure. And when I make this observation, I do not think of a principium contradictionis.

This quotation does not show that Brouwer objects to the validity of logic or, still less, to the meaning of logical constants. It makes it clear, however, that the logical constants, logical rules, and principles are for him derivative from some operations of intuitive mathematics.

Given this, what sense can Brouwer give to the notion of proof? In particular, what is his assessment of the axiomatic-deductive method? The first axiomatic and deductive mathematical theory, which moreover set the pattern of what the axiomatic-deductive method is, was that of Euclid's *Elements*. We shall thus focus on it and Brouwer's appraisal of it. Euclid's work consists of thirteen books, each starting with a set of definitions, a list of postulates, and a list of 'common notions', that is, axioms. Definitions introduce new concepts that are explained in ordinary language; these can hardly be viewed as formal definitions given in terms of the primarily assumed primitive terms. Axioms describe basic properties of numbers and geometrical objects; as an example one may take the famous claim "the whole is greater than the part". Finally, postulates delineate what basic constructions are possible. Now, the essential feature of this system is that all theorems are to be *derived* from the accepted definitions, axioms, and postulates. However, we should not be misled into thinking that derivations in the *Elements* are logical inferences of new statements from precisely formulated axioms by means of explicitly stated syntactic rules of inference. Clearly, the language of the *Elements* is not formalized. Definitions, axioms, and postulates are sufficiently exact as informal explanations or statements of some truths but they are formulated in ordinary language. Subsequent studies of Euclid's proofs have shown that some of them appeal to intuitively evident but tacitly assumed premises. Still further, it is not so certain in what sense Euclid's proofs appeal to logical means. It is true that some sentences that occur in their formulation are built with expressions marking logical constants, so the sentences have a form of, say, implication sentences, or disjunctive sentences. However, since it is hard to conceive of a mathematical theorem being formulated without the help of those expressions, the mere presence of logical constants in a formulation of a proof does not testify to its being logically deduced. To take an example, in some proofs (e.g., of Proposition 6, Book I) we may discern a figure that looks like an application of the double negation elimination. We read there:

> Therefore *AB* is not unequal to *AC*; it is therefore equal to it,[28]

where *AB* and *AC* are sides of a triangle that subtend the equal angles and the proof establishes that in a triangle of two equal angles, the sides which subtend the equal angles are equal to one another. In the same proof we encounter an appeal to impossibility or absurdity, as it is found that a smaller triangle is equal to a greater one, and we get an additional hint that it is an *impossibility* by learning the number of a postulate that conflicts with the equality of these triangles. Nevertheless, despite such appeals to logical notions and principles, there is an inclination to say that they are not essential. The proofs are presented in a way that should enable the reader to effect a certain geometrical construction. They guide him, so to speak, by showing what moves should be effected and in which order. They also provide justification of why the move is possible by pointing to an appropriate axiom or postulate. And axioms and postulates are not conceived of as arbitrary restrictions on the possibility of geometrical constructions, but rather as evident and intuitively known truths. In this interpretation proofs are thus conceived of as instructions, or ways of convincing someone of the truth of their conclusions.

The other approach sees in Euclid's *Elements* an outline, or a yet-imperfect realization, of the axiomatic theory that was subsequently elaborated and improved on. In this interpretation, the issue of the evident character of axioms and postulates is to a large extent ignored. Instead, it is the task of formulating precisely the axioms and postulates and disclosing the premises tacitly assumed in proofs that gains in importance. Proof is identified with deduction of a sentence from the assumed axioms and postulates by means of rules of inference and laws of logic. Validity of a given proof is evaluated by examining whether all its inferences are made in accord with the assumed rules.

Now, it is obviously this latter interpretation of *Elements* that caught on among mathematicians in the subsequent centuries, as they were preoccupied with issues like independence of axioms, or making clear which axioms and postulates are appealed to in any given proof. From Brouwer's perspective, this carried a danger of focusing the attention entirely on the language of mathematics, that is on operations of deriving new sentences from those already asserted by means of some rules of transition, without giving a thought to whether the sentences as well as their transformations represent anything in intuitive mathematics. The danger materialized in investigations of the fifth postulate of the first book of *Elements*, when it turned out that replacing this postulate by its negation did not lead, as expected, to the emergence of a contradiction, though it allows for deducing some odd and counterintuitive theorems. In this way non-Euclidean geometries emerged whose theorems apparently defied what were believed to be properties of real space. Their discovery opened the way to completely ignoring the intuitive elements in mathematics. Given a system of axioms and a consistent set of rules of inference, a mathematician might set about the task of deducing new sentences, without bothering about what truths of intuitive mathematics these sentences express or what constructions of intuitive mathematics are represented by deductions. Now, it is precisely this way of doing mathematics that Brouwer objects to. While speculating over how Euclid might have thought of his system, he comes to the following alternative. On the one hand, Euclid might have thought of his axioms as describing a certain mathematical structure, say, Euclidean geometry, and understood his mathematical activity as operations on its substructures. His proofs, as they occur in *Elements*, do not have mathematical significance, as these are merely linguistic structures accompanying the mathematician's exploration of the structure, and means of convincing others of his line of thought. On the other hand, as Brouwer puts it, Euclid might have

(...) lapsed into the mistake which so many people made, thinking that they could reason logically about other subjects than mathematical structures built by themselves (...) (1907, p. 76)

Brouwer's reservations to using the axiomatic-deductive method in mathematics stem from two sources. First, he fears that logical deductions may lead one too far, as they may allow the deduction from intuitively true axioms of propositions that do not correspond to any intuitive constructions. He also sees in the axiomatic system an unnecessary restriction artificially barring some constructions that may be achieved in intuitive mathematics. Secondly, he objects to the idea, which is somewhat linked to the discussed method, though not inherent in it, that in logical derivations one may ignore the meaning that the sentences involved have in intuitive mathematics. In fact, neither reason tells against the axiomatization of a constructive mathematical theory, under the condition however, that the constructions of this theory have already been effected by the methods acceptable to the intuitionist. And although Brouwer himself never took an interest in axiomatizing any part of intuitionistic mathematics, his attitude towards axiomatization grew more lenient. For instance, he encouraged Heyting, his student at the time, to "make the axiomatics of projective geometry intuitionistic".[29] This lenient view on axiomatization was later adopted and further elaborated by Heyting.

3.4. *What Is Logic?*

It is understandable that an attempt to formalize mathematics does not make much sense for somebody who holds mathematics to be sharply separated from its language. What is less understandable is Brouwer's opposition to the use of logical tools (logical laws and principles or rules of inference) as means of carrying out mathematical constructions. Some rationale can be found in his reversal of the relation usually believed to hold between logic and mathematics. In his doctoral dissertation he states the issue very clearly:

(...) mathematics is independent of *logic*, logic does depend upon mathematics. (1907, p. 73)

To appreciate this claim fully, we need to reflect on what he understands by logic, or logical laws. Importantly, he states explicitly that the latter are "laws of reasoning or of human thought" (*ibid*, p.72). However, as he never gave a full list of claims that he considered laws of logic, it is hard to say what logic he had in mind. The laws that are discussed in his papers comprise the principle of non-contradiction, the principle of the excluded middle, double negation elimination, syllogism, the law of the distribution of conjunction and alternative, the principle of identity, and de Morgan's laws. On the other hand, we may safely assume that he was not familiar with works of the founders of modern logic, like Frege or Boole. He never mentions Frege in his papers; moreover the prevailing opinion has it that Brouwer was not familiar with Frege's concept of logic and details of his program of logicism. Similarly, at the time he launched an attack on classical logic, he rather did not know Boole's *Laws of Thought*, although his much later paper (1955) is devoted to discussing relations of intuitionism to Boole's logic. Boole's name is mentioned there together with Jevons, Peirce, and Schröder. Symptomatically, he still interprets classical logic there as yielding "(...) a formal image of the laws of common-sensical thought." (*ibid*., p. 551)

Thus, by 'classical logic' he presumably understood a body of laws of propositional logic with, perhaps, the theory of syllogism. He distinguishes this subject from Peano's and Russell's symbolic logic that he considers to be based on the notion of *propositional function*, that is

statement about *x*, or *x* and *y*, more generally about a certain number of variables, in which *every* substitution for these variables is allowed. (1907, p. 88)

Now, Brouwer's first idea is that laws of classical logic, if they are true, represent in language mathematical propositions of high generality, namely those dealing with relations of parts vs. the whole. One may consider, for instance, the following reasoning, which he mistakenly calls a syllogism:[30]

All men are mortal
Socrates is a man,   *ergo*:
Socrates is mortal.   (1907, p. 74)

Brouwer argues that the intuitive reasoning lying behind the above figure consists first in projecting into the world of perception a mathematical system which consists of a finite set of objects, and then linking some objects of this set to element(s) of another set, namely a set of predicates. In the above case, the set of all objects contains, among others, Socrates. The set of predicates contains (among others) the predicates *man* and *mortal*. Now, the elements of the set of objects that are linked to the predicate *man* are contained in the set of elements linked to the predicate *mortal*—that is what the first premise states. Socrates, an element of the set of objects that is linked to the predicate *man*, is also linked to the predicate *mortal* and this is expressed by saying that Socrates is mortal. Thus, what corresponds to a linguistic regularity of a syllogism is a comparatively simple mathematical construction.

Contrary to this picture, the logician who contemplates an instance of syllogism does not think of a mathematical construction. He looks at words and realizes a surprising linguistic regularity: whenever two *sentences* of certain forms are accepted, a third sentence, of such and such form, can also be accepted. Now, if these regularities themselves are studied by mathematical means, they are elevated to aprioristic *laws* of logic. As a result one may be inclined, while contemplating an instance of such a law, to ignore an underlying mathematical construction that usually relates to the relation of parts to whole. In Brouwer's view, logic comes into being when mathematicians have recourse to language in order to keep track of mathematical constructions and communicate them to others. At first, linguistic structures serve to report on the effected constructions; in this sense they are said to accompany mathematical reasoning. Moreover, in the language of those reports some special expressions occur that allow for the formation of composite sentences, like for instance *if ... then, and, or, not* etc. These reports on mathematical constructions exhibit certain regularities. In the next phase, that language itself is subjected to mathematical attention and as a result linguistic regularities are conceived of as 'laws' of a newly postulated mathematical system. A mathematical system projected on a variety of mathematical sentences in this way is, according to Brouwer, the subject of logic. This means that what was primarily only a regularity between various types of asserted sentences is elevated to a law, which is supposed to govern every reasoning of the appropriate sort. The empirically given fact that we can rely on a passage from an assertion of a certain type to an assertion of another type is now supposedly justified by the assumption that such regularities are instances of an *a priori* law. This justification ignores the fact that a linguistic regularity like syllogism holds only because one can create a respective mathematical construction that relies on the transitivity of the relation between part and whole. One consequence of such rather empirical views of the laws of logic as somewhat idealized regularities occurring in linguistic reports on reasoning is that those laws are made dependent on a milieu of language-users. Firstly, it is up to a given society how deeply mathematical constructions are analyzed, that is, how far it goes in discerning sub-structures of the constructions. Secondly, which elements of mathematical constructions have names also depends on the milieu. From this Brouwer draws the conclusion that

(...) given the same organization of the human intellect and consequently, the same mathematics, a different language would have been formed, into which the language of logical reasoning, well known to us, would not fit. (1907, p. 74)

And he goes even further, as he speculates whether or not this is so in the case of the language of a community isolated from European culture.

One consequence of these views is that differences between languages may result in differences between theoretical logics, as the latter result from subjecting a language to mathematical treatment. What is more, these latter differences are traceable rather to the organization or ways of life of a given community than to alleged differences in human intellects. It follows that

We infer that theoretical logic as well as logistic are *empirical sciences* and that they *apply* mathematics; consequently they can yield no information whatsoever on the organization of the human intellect; there would be better reasons to reckon them under *ethnography* than under *psychology*. (1907, p. 74)

At first sight the conclusion that logic is an empirical subject, quite similar to ethnography, may strike one as absurd.[31] The absurdity is felt especially strongly by someone who under Brouwer's notion of logic mistakenly understands contemporary mathematical logic, which is hard to differentiate non-arbitrarily from mathematics. Thus, to grasp Brouwer's discovery it is necessary to think of its context, which in our opinion was provided by a tradition, quite powerful in the nineteenth century, of interpreting laws of logic as describing, or perhaps prescribing, which intuitive reasonings are correct. Given this context, Brouwer rightly observes that if these laws relate at all to reasoning, they do so only indirectly, by characterizing which schemata of arguments, *as expressed in language*, are valid. However, he does not agree that our intuitive reasoning can be fully and adequately expressed in language. The idea that a language of thought and reasoning can be designed is more of a postulate than a reality. As Brouwer was rather pessimistic about its realization, he was left with actual languages: ethnic ones, professional jargons, dialects. These languages, if used to express intuitive reasoning, exhibit some regularities that obtain between various kinds of assertions; such regularities, if they are sufficiently idealized, are identified with laws of logic. Now, if one is only attracted to this line of thought, and moreover, is prone to the somewhat suspicious idea that there might be different languages in which intuitive reasoning is expressed in forms very different from those known to occur in our own language, then that person would be inclined to think of logic as relative to natural language.

3.5. *Mathematics vs. Logic*

We said above that, according to Brouwer, laws of logic emerge as an effect of using language in order to express intuitive reasoning. If the reasoning is mathematical, it is *logistic*, as he calls it, presumably referring to Peano's system or Russell's logic from the *Principles of Mathematics*. It emerges from mathematical investigations of the language that expresses mathematical constructions. Brouwer insists that the idea of subsuming the language of mathematics under mathematical investigations is rather recent, having taken hold only in the last decades of the nineteenth century, although it had been anticipated by Leibniz (1907, p. 74). To investigate further the relation that Brouwer assumes to hold between logic and mathematics, let us start with a platitude, namely, that in reports on mathematical constructions, mathematicians have recourse to language containing such expressions as 'if ... then', 'and', 'or', 'not', 'every' which hint at what today we call logical connectives and quantifiers. From his perspective, it would be better if mathematicians managed to communicate without using such expressions. As he suggests to the supervisor of his doctoral thesis, Korteweg, it is the use of language containing logical expressions that generates paradoxes of set theory; they would disappear if mathematicians found a 'pure language' that did not contain such troublesome expressions. Thus, he sees in the existence of these expressions an imperfection of language, something that points to its 'poverty'. In a letter to Korteweg from Jan. 23, 1907,[32] while explaining a chapter of his dissertation, he states:

At the beginning of the chapter I show that mathematical reasoning is *not* logical reasoning, that only because of the poverty of language it makes use of the connectives of logical reasoning (...) (VS, p. 503)

However, since mathematicians do not have at their disposal a pure language that does not contain logical connectives, both the mathematician and the logically minded use the same language. Nevertheless, the difference between the two approaches shows up in the distinct ways they construe sentences containing logical constants. Let us follow one of Brouwer's examples that points out what construals he has in mind. While explaining this issue to Korteweg he writes:

The theorem: *if a triangle is isosceles then it is acute angled* is presented as a logical theorem and the predicate 'isosceles' for triangles is taken to imply the predicate 'acute angled', i.e. one imagines all triangles (e.g. of a plane) represented by points of an R^6 and then one observes that the region representing all isosceles triangles is contained in that which represent all acute angled triangles. This really holds; here the logical formulation and therefore the logical language can safely be used. But the mathematician who formulates the above theorem, due to poverty of language, as a logical theorem, thinks of something different from the above logical interpretation. He conceives of a construction of an isosceles triangle and finds that after completing the construction the angles turn out to be acute, or that postulating a right or obtuse angle the construction fails. In other words he conceives the theorem in a mathematical, not in a logical interpretation. (VS, p. 503)

Let us start with the implication sentence: 'if a triangle is isosceles then it is acute angled'. While interpreting it, a logician is oriented towards language. He discerns in it a phrase pointing to implications and some predicates. The predicates, according to him, define their extensions, that is, classes of objects that fall under them. He thus thinks of classes of all triangles, isosceles triangles, acute angled triangles. Finally, he construes the theorem as a claim that the class of all isosceles is contained in the class of all acute angled triangles. When mathematically construed, on the contrary, the sentence is a claim about the possibility of effecting a certain construction that shows that any constructed isosceles has all its angles acute. One may naturally wonder what is wrong, according to Brouwer, with presentation of theorems as 'logical' statements. Although the logical interpretation is allowable in the example discussed, because one may effect the mathematical construction demanded by mathematical interpretation of that theorem, the suggestion is that in some other cases, an appropriate mathematical construction may be not known. We may approach this issue by first concentrating on logical reasoning about the external world. In Brouwer's philosophy, a necessary precondition of such reasoning is taking a mathematical view of sensations, that is, arranging them in some mathematical structure. Language is then used to talk about mathematical constructs that represent aggregates of sensations. Now, as Brouwer believes that the mathematics of the relation of part to whole is trivial, logical reasoning about the external world, *if applicable*, yields only trivial results, which could be read off from its assumptions without the need of having recourse to deductions. He claims, moreover, that if a logical reasoning leads to acceptance of a sentence which was not obvious from the beginning, this in itself is a reason to deny the validity of this reasoning! (VS, p. 504) In this case his suspicion is that the part of the external world to which the reasoning relates is not mathematically manageable, so logic was not applicable there. What is involved, however, in the claim that a part of external world is not mathematically manageable? The explanation that one finds in the letter is hardly convincing, since the feature in question is alleged to follow from the fact that:

(..) one does not believe any more (...) that the world consists of a very large, but finite number of atoms, and that every word must represent a (therefore also finite) group or group of groups of those atoms. (VS, p. 504)

It thus seems that what precludes the external world from being represented by a mathematical system is the infinite number of objects. Moreover, the same diagnosis applies to logical reasoning as that applied to mathematics: it can safely be used if a domain of mathematical objects is finite, but otherwise not. According to Brouwer, the distinction between trivial but reliable logical reasoning on the one hand, and non-trivial but unreliable reasoning on the other, occurs in mathematics as well. Thus, there are 'theorems' like

If a number is prime, then it is natural,

which are logically justified by constructing a 'mathematical system', i.e., the set of all primes and the set of all natural numbers, and noticing that the former is contained in the latter. This justification is, however, otiose since it is clear that every construction of a prime presupposes that it is a natural number. The other possibilities are statements like

The set of all denumerable ordinals is non-denumerable,

that have proofs that proceed in accord with the rules of logic, but refer to mathematical systems they are not constructible according to Brouwer's standards. He maintains that the set of all denumerable ordinals is a logical entity only, since although it can be introduced by defining a class that satisfies some predicate, it cannot be constructed by means of the intuition of two-ity. The logical rules and principles involved in the deduction of this statement are unreliable, as they permit the assertion of sentences to which nothing corresponds in mathematical reality.

Thus, we have the following teaching on the relation between logic and mathematics. One may say that mathematics deals with deductions of mathematical theorems that lead from some sentences, possibly containing logical constants, via some rules, to other sentences, again possibly with logical connectives. Nevertheless, one needs to remember that logical principles and rules are in fact disguised theorems concerning the relation of parts to whole. Thus, deductions are reliable as far as the mathematical systems they pertain to are constructible. If this is so, however, the deductions are hardly needed as their conclusions can be derived from the assumptions and the mathematics of part and whole. On the other hand, deductions whose constituent sentences or formulae do not pertain to constructible mathematical systems are not reliable as they allow one to assert theorems of no mathematical meaning.

3.6. *Against* Begriffe

Brouwer's idea of language as a vehicle of will-transmission has one more significant consequence: it leads to a change in the understanding of the concept of truth. The crucial part in this argument is played by the rejection of *concepts* or *Begriffe* as he calls them. For the sake of the argument, let us assume with Brouwer that what we call *language* is in fact a complicated system of gestures and signs by which humans convey commands, express wishes, ask for help, and attempt to bring about desired mental states in others. Utterances in that language are made in accord with laws governing logical connectives (*Aussagenverknüpfungsgesätze*). (1929, p. 423) These laws allow one correctly to reason about facts of the external world, given that the number of its elements is finite. From sentences that express a fact obtaining in the external world, one may derive with the help of the principles mentioned some other sentence(s). However, the appeal to logical laws is otiose, since the conclusions are intuitively perceived as true before any logic is applied. As we have mentioned, the success of the deductive method, as applied to the external world, is due to the projection of a *finite* mathematical system on objects and facts of the external world. That system consists of a finite number of discrete elements located in space-time and linked by a finite number of relations. (*ibid.*, p. 423) Despite the success of this linguistic method, however, it should not be forgotten that language, according to Brouwer, has no aspects other than being a means of communication by invoking in other persons' minds appropriate images or, as he prefers it, of transmitting will. Brouwer goes on to say that, as a result of a persistent prejudice in favor of the conception of an ideal language of thought and reasoning, the insight that transmitting will is the sole aspect of language has never caught on. The prejudice consists in the belief that words, apart from being means of invoking in other minds appropriate images, have another function, which is pointing at what he calls fetish concepts (*fetischen Begriffe*) and their complexes. What are then these concepts and their relations? Brouwer explains that these are believed to exist independently of human causal attention, and moreover, that it is a part of this misconception that "logical principles present aprioristic laws that govern concepts and their complexes." (*ibid.*, p. 423) The mention of the causal attention suggests that what differentiates concepts from the objects allowable in Brouwer's philosophy is that the former cannot be perceived, or given to the mind. What differentiates a concept that is allegedly linked to a word from the image that this word invokes in somebody's mind is that the image, if invoked, must be perceived, while the concept may transcend our capacity to perceive it intuitively. Brouwer's remarks are nevertheless too scanty to form a definite opinion on what he understands by concepts. On the one hand, it may be argued that he is inveighing against Fregean senses; these, however, are by their very nature comprehensible by any competent language-user. Secondly, he may be interpreted as attacking the idea that words *refer* since, in his view, they only invoke images. The rejection of a claim that a word may refer to an object even though it may be impossible effectively to recognize which object it refers to, accords reasonably well with Brouwer's metaphysical views. Finally and perhaps most plausibly, one may argue that what he calls concepts are extensions of predicates. If so, his assault on *Begriffe* should be seen as part of a wider strategy of objecting to a linguistic means of introducing objects—'linguistic' because it starts by announcing a predicate and then postulates the class of all objects that fall under it, ignoring completely the need to have them given intuitively.

A second ingredient of the doctrine of concepts under attack identifies some compositions of concepts with truths, where some of these truths are only of an ideal nature, as they cannot be tested. Let us introduce after Brouwer the following threefold distinction of truths. To begin with, we have evident truth, i.e., a compositions of concepts that points to an undeniable (*unleugbare*) fact. As I interpret it, evident truth is a true statement for which one has adequate evidence on a given occasion. Secondly, we have controllable sentences (*controlierbare Aussagen*), the truth or falsity of which can be tested in given circumstances. Thus, although evidence for the controllable sentence may be missing on a given occasion, the subject has a method to decide whether or not the sentence is true. Finally, there are ideal truths (*ideale Wahrheiten*) that are believed to be true, though their truth is not, or cannot, be seen intuitively (*ibid.*). To get a grip on the last notion, we need to reflect on how one arrives at ideal truths. Given an evident truth, that is a complex of concepts to which an appropriate mathematical construction or complex of sensations corresponds, it is believed that principles of logic permit derivations of new, evident truths. However, according to Brouwer, they also make it possible to derive from evident truths other truths that are not evident. One principle responsible for such derivations is identified by Brouwer with double negation elimination. In such a case a derivation starts with evident premises and then draws a conclusion on the grounds of the contradictority of its negation. It may happen that a statement is accepted only on the grounds of an indirect argument, as its direct demonstration is not known, and may never be known. In such a case, the statement expresses an ideal truth. The classical mathematician nevertheless asserts p when he has grounds to assert only not-not-p. Another example of means that, according to Brouwer, permit the introduction of 'fetishes' of ideal truths is the comprehension axiom (*ibid.*, p. 424). These examples illustrate in fact how Brouwer's conception of evidence must be understood. In the case of mathematical statement, evidence is identified with an appropriate intuitionistic mathematical construction, where evidence for not-not-*p* importantly does not count as evidence for *p*. Theorems pertaining to cardinalities higher than denumerable are ideal truths, since what they state cannot be constructed by following the intuition of two-ity.

In his criticism of ideal truths Brouwer seems to go even further, as he hints at the possibility that an application of principles of logic to ideal truths may lead to contradictory results.[33] In the course of a logical deduction a contradiction may result from either the assumed premises, or from the accepted rules of inference. Now, in the case of many proofs of classical mathematics, a proof proceeds *via* ideal truths, that is those to which no intuitionistic construction may correspond. The classical mathematician nevertheless applies logical rules and principles to sentences expressing such ideal truths. Such derivations sometimes end up with a paradox, as was the case in set theory. But, as Brouwer observes, the laws of logic are so undisputed and belief in ideal truths so well-entrenched, that whenever a contradiction occurs in the course of a deduction, people always assume that the premises are incorrect and never entertain any doubts about the validity of the laws of logic, or their application to ideal truths. (1929, p. 424)

3.7. *What Is Truth?*

If truth is not a feature of statements, then what is true? In Brouwer's view, truth refers to extra-linguistic entities, for instance to mathematical construction, complexes of sensation, and the like. But can thoughts be said to be true? Brouwer assents but only in so far as the thoughts are 'experienced'. As he puts it

(...) truth is only in reality i.e. in the present and past experience of the consciousness. Amongst these are things, qualities of things, emotions, rules (state rules, cooperation rules, game rules) and deeds (material deeds, deeds of thought, mathematical deeds). But expected experiences, and experience attributed to others are true only as anticipations and hypotheses; in their content there is no truth. (1948, p. 488)[34]

Brouwer's concept of truth relies on two presuppositions. First, truth is an ontic property, which means that whatever is, is true. Thus, the word 'true' is synonymous with 'real'. For instance, an object is true if it really is, an event is true if it really happens, a process is true provided it really takes place. But secondly, the meaning of the expression 'there is something' is limited, since 'to be' means, more or less, 'to be experienced'. Thus, objects and qualities are true as far as they are perceived; mathematical constructions are true but only if they are carried out; and similarly, rules and deeds are true only if the subject is aware of them. What Brouwer says about the truth of 'expected experiences' is illustrative: as the subject is consciously aware of such expectations, they are true but only as *expectations*. Their content, which is about an 'unconstructed' object, quality or something else, is not true.

The idea that 'there are no non-experienced' truths follows from the rejection of the view that something objective and unknown to us may guarantee that a statement is true, and accordingly, that a statement can be true, although what makes it true escapes human recognition. Against realism Brouwer advocates a 'metaphysical constructivism' that sees in everyday and scientific objects creations of the mind out of sensations. Also, he rejects the concept of a super-subject who experiences all sensations and knows all the possible mathematical constructions, and who guarantees that a construction is true, independently of whether or not it has been carried out by a human being. Similarly, he discards the objective sphere of meaning, concepts, and *ideal truths* that are supposedly expressed by means of language.

A more detailed analysis shows that the following theses, which are inherent to any objectivist conception of truth, are rejected:

1. that truth exists "independently of human thought and is expressible by means of sentences called 'true assertions'." These true assertions assign "certain properties to certain objects or state that certain objects possessing certain properties exist or that certain phenomena behave according to certain laws" (1955, p. 551);
2. that one may extend the stock of true assertions by means of the laws of logic, which permit one to deduce further evident assertions from an initial stock of evidently true assertions, called *axioms* (*ibid.*, p. 551);
3. that if the term 'false' denotes what is 'converse to true', "each assignment of a property to an object or of a behavior to a phenomenon, is either true or false independently of human beings knowing about this falsehood or truth, so that e.g. contradictority of falsehood would imply truth (...)" (*ibid.*, p. 551);
4. that any mathematical assertion is either true or false, independently of whether anybody knows it; moreover, if the human race became extinct, mathematical truths would survive.[35]

These four statements are rejected as forming the core of a confused conception of truth.

On may suspect that Brouwer's doctrine of ontic truth, with its slogan 'no non-experienced truth' brings the concept of truth close to being redundant. In a reported discussion of the signific circle, when asked to explain his account of truth in mathematics, Brouwer went on to say that

> (...) truth is a general emotional phenomenon, which by way of accompanying appearance (Begleiterscheinung) can be coupled or not with formalistic studies of mathematics.
> (1937, p. 451)

The passage clearly shows that whereas truth is only an emotional or psychic phenomenon, it may nevertheless be 'coupled' with some structures of a formal language of mathematics. Thus, we may risk the supposition that statements are indeed capable of being true, even though truth refers first and above all to extra-linguistic structures: objects, properties, rules etc. On this construal, the statement is true only if it expresses (or points to) a fact that is experienced by the subject. Let us repeat, however, that to require that a statement be true only if the appropriate construction has actually been carried out makes the concept of truth entirely subjectivist and time-dependent (see section 2.4). For this reason it is better to relax the strict requirement of actual construction and say that the statement is true provided it is known how to effect the desired construction.

3.8. *The Validity of Laws of Logic*

3.8.1. *Weak Counterexamples*

The doctrine of 'no non-experienced truth' and the picture of logic as a science of linguistic regularities bear on the question of the validity of laws of logic. In the framework of Brouwer's intuitionism this question takes an 'empirical' form. It translates into the query as to whether among various regularities of linguistic structures that 'accompany' mathematical constructions carried out by the subject are those representing the discussed laws of logic. It is illuminating to take a look at Brouwer's own formulation of the problem. He asks:

> Will human beings with an unlimited memory, while surveying the strings of their affirmation in a language which they use for an abbreviated registration of their constructions, come across the linguistic images of the logical principles in all their mathematical transformations? (1933, p. 443)

Before asking this question he makes two idealizations. First, he assumes that the memory is unlimited in its capacities, which in Brouwer's view guarantees that constructions effected by the subject are error-free. Secondly, it is assumed that the symbols of language correspond univocally to elements of mathematical mental constructions. The question requires us to survey all the strings of symbols, as they record processes of creating various constructions, and find those sequences of symbols that represent laws of logic. It is somehow unclear how the signs of logical connectives may appear in such reports on the 'languageless' mathematical activities of the subject. However, recollecting Brouwer's remarks on negation, it is understandable how at least the sign of negation occurs in the reports. As he told us, while a logician talks about negation of a statement, a mathematician perceives that an attempted construction corresponding to that statement 'no longer goes' (1907, p. 72). In other papers he explains negation by linking it to a construction that leads to a stop or absurdity, or runs into the impossibility of its being fitted into another construction. Thus, we may say that the negated statement $\neg p$ corresponds to a construction which shows the impossibility of any construction corresponding to p. What, then, corresponds to a double-negated statement $\neg\neg p$? Clearly, a construction showing the impossibility of the impossibility of any construction corresponding to p. Thus, $\neg\neg p$ is asserted only if a proof that $\neg p$ has led to an absurdity.

Is the principle of double negation elimination then valid? For this to be the case, any 'string' of the form $\neg\neg p$ should be replaceable by p. That would mean that any proof of the impossibility of the impossibility of p should be capable of transformation into a direct proof of p. As there is no known method of transforming indirect proofs into direct proofs, not every string of the form $\neg\neg p$ is replaceable by p, and this, according to Brouwer's standards, shows that the rule of double negation elimination is not a valid rule of reasoning. If we look at this from the perspective of the natural deduction formalization of logic, Brouwer's argument is readily interpreted as demanding the rejection of the rule of double negation, i.e.,

$$\frac{\neg\neg p}{p}$$

Now, as is well known, the differences between the classical and intuitionistic systems of natural deductions for predicate logic stem from the fact that the former assumes the above rule, while the latter does not. Clearly, the rejection of the rule has an effect on what can be asserted, in particular, on assertability of instances of the principle of the excluded middle. With the rule not admitted, $p \vee \neg p$ cannot be asserted merely on the grounds of the contradictoriness of $\neg(p \vee \neg p)$. We may thus trace the objections to the validity of the principle of the excluded middle to the rejection of the rule of double negation elimination. Nevertheless, as Brouwer's papers document, his main obsession was the rejection of the principle of the excluded middle, and not the rule discussed. For this reason, we need to analyze his original objections to the principle. Brouwer assumes that $p \vee q$ can be asserted only if either a construction corresponding to p, or a construction showing that q, or both constructions can be carried out. This leads at once to the conclusion that $p \vee \neg p$ may only be asserted if one of the two constructions: that p or that $\neg p$ can be effected. And, as there are statements of which it is neither known how to prove them, nor how to derive a contradiction from the assumption that they hold, counterexamples to the principle so interpreted abound. Similarly, a statement that at present can only be proved by indirect means, that is by deducing a contradiction from its negation, serves for Brouwer as a counterexample to the principle of double negation elimination. These counterexamples are usually referred to as 'weak' since they attempt only to show that some instances of these principles cannot be asserted in the present state of knowledge, though the principles are not contradictory.

Importantly, however, the principle of the excluded middle has here received a substantial meaning. It may now be expressed as follows:

An assertion A can be *judged*, i.e. can either be proved to be true or be proved to be contradictory. (1954, p. 524)

Clearly, this is a substantial claim, much stronger than the classical principle stating that A or not-A is true, where truth is understood as a property that statements determinately possess or not, no matter whether or not anybody has a means of recognizing it. But as we have already seen, Brouwer rejects the objective concept of truth, and accordingly the classical interpretation of that principle has little meaning for him. Thus, a crucial part of the intuitionistic critique of classical logic is the controversy over truth; we shall investigate later what rationale Brouwer may have for rejecting objective, bivalent truth.

The rebuttal of bivalence, that is the claim that any statement is determinately either true or false, drives Brouwer further to replace the twofold distinction between truth and falsity by four clauses:

(1) a has been proved to be true; (2) a has been proved to be false, i.e. absurd; (3) a has neither been proved to be true nor to be absurd, but an algorithm is known leading to a decision either that a is true or that a is absurd; 4) a has neither been proved to be true nor to be absurd, nor do we know an algorithm leading to the statement either that a is true or that a is absurd (1955, p. 552),

where a is a mathematical statement. As clauses (1)–(3) can be summarized by saying that it can be decided whether or not a holds and that this is also the meaning of the principle of the excluded middle that the intuitionist proposes, it is clause (4) that matters. To argue that it is (4) that makes the principle of the excluded middle not valid is to assume that it is not reducible to (3). But, as Brouwer explicitly states, the division between statements of (4) class and those of the others is not permanent, as a problem not decided at a given time may later be shown to be true, or reduced to absurdity. Also, one cannot claim that among statements of the (4) class *there are* any that can never be solved because of mathematical reasons only. Due to the intuitionistic understanding of negation, this positive claim (if at all meaningful) that there *are* truly undecidable statements amounts to a contradiction. Thus, we must understand the thesis that clause (4) is not reducible to the other clauses in a rather delicate manner. We should say that we refuse to believe optimistically that all problems are solvable, although on the other hand we must abstain from declaring that there are truly undecidable problems.

3.8.2. *A Reconstruction of Brouwer's Argument*
We have been contemplating Brouwer's argument against the validity of the principle of the excluded middle, which draws heavily on picturing logical laws as idealized regularities obtaining in the reports on the mathematical activities of the subject. This make the issue of the validity of a law or a rule of logic tantamount to checking whether corresponding sequences of symbols may occur in reports on mathematical constructions. According to our suggestion, the argument tries to establish, on the assumption of the restricted concept of evidence, that the rule of double negation elimination is not universally valid, and then derives from this fact a change of assertion conditions for disjunctive statements. As to Brouwer's own examples, they appear to rely on his rebuttal of the classical concept of truth. Owing to this rebuttal, he renounces as meaningless the classical mathematician's supposition that

each instance of the excluded middle is *true*. Thus, to become converted to intuitionism, it is necessary to find an interpretation of Brouwer's arguments against the classical understanding of laws of logic that does not hinge on his rather controversial views on logic and language, or to find a convincing rationale for abandoning bivalent truth.

Let us start with reconstructing such an argument. A motivation for attempting this sort of interpretation can be found in his (1929) paper, in which Brouwer introduced a distinction between evident statements, controllable statements, and ideal truths (*ibid.*, p. 423). To repeat, a statement is evident if one has evidence for its assertion, and is controllable in a given state of knowledge if one can find in this state of knowledge evidence either for its assertion or for assertion of its negation. In the case of mathematical statements, evidence is identified with possessing the appropriate intuitionist mathematical construction. Brouwer's accusation against classical logic may be summed up by saying that it permits the derivation from evident statements of ideal truths, that is, statements for which evidence is not known in a given epistemic situation (state of knowledge). It is necessary to think of evidence as largely independent of logical rules of inference. Evidence is not something that is arrived at by appeal to rules governing transitions between sentences. On the contrary, these rules stand in need of justification: only those that preserve evidence for assertions are valid. Consequently, in the case of transition from one sentence to the other that accords with a valid law of logic, one should be in a position to obtain the evidence for the latter sentence from the evidence for the former. Let us illustrate these ideas by considering an instance of the rule of double negation elimination.

Suppose I have been told by a reliable informant that in a town, say Sucha, there is one hairdresser's. Trying to locate it, I have made an investigation and found that on every street of Sucha, with the sole exception of Lubon street which I have not walked through, there is no hairdresser store. By relying on my informant, eliminating possibilities that a hairdresser's is on this or that street, and finally realizing that there is only one remaining possibility, namely Lubon street, that I have not searched, I have procured the evidence for asserting the sentence:

It is not the case that there is no hairdresser's on Lubon street.

The question is whether this evidence allows the assertion of the sentence without double negation, namely

There is a hairdresser's on Lubon street.

What first comes to mind as a typical candidate for evidence for the latter sentence is seeing or remembering a hairdresser's on Lubon street. If so, there is some difference between the two sorts of evidence, that is knowing, by eliminating all other possibilities, that the hairdresser's is on Lubon street and seeing or remembering the hairdresser's on Lubon street. The assumption that the former information serves as evidence for the two sentences above tends to blur the difference between the two sorts of evidence. Thus, if one prefers that the difference not disappear, he had better accept that the evidence for a double negated sentence does not *automatically* count as evidence for the sentence obtained from the former by crossing out the double negation.

In our case, however, of a hairdresser's being located on Lubon street in Sucha, the distinction between the two kinds of evidence shrinks to triviality. This is so because, after all, if I know that the hairdresser's is not located on any other street than Lubon street, I may, if I wish, walk through this street and see it there (on the presumption that the informant was reliable and I was not mistaken in my findings about all the other streets). Although I do not see or remember seeing the hairdresser's on Lubon street, by thoroughly examining this street I will find it there. Thus, we may assert that there is a hairdresser's on Lubon street merely on the grounds of the information provided by the informant and by eliminating all the other possibilities. In this case the rule of double negation elimination can be applied safely. Nevertheless, the example shows that the validity of the rule seems to be a contingent matter since, if we allow for undecided sentences (i.e., such that in a given epistemic situation it is neither known how to procure evidence for the assertion of that sentence nor for the assertion of its negation), the evidence for a double negated sentence will not help us to obtain the evidence for the sentence with eliminated double negation. To illustrate this, let us focus our attention on Heyting's σ-example. The instruction for the construction of σ is as follows:

Consider the occurrence of the series $1, 2, 3, 4, 5, 6, 7, 8, 9$ in the decimal expansion of π. Now, if the last 9 of the series $1, 2, 3, 4, 5, 6, 7, 8, 9$ does not occur in the n-th place of the expansion of π, put down 3 in the n-th place of the decimal representation of σ. If such a 9 occurs in the n-the place of π, write 0 in the n-th place of the expansion of σ and stop developing the expansion. (Heyting 1956, pp. 17–18)

On the grounds that a contradiction follows from the assumption that σ is not rational, we may assert the sentence

It is not the case that σ is not rational.

But does this sort of evidence help us obtain the evidence for the following sentence?

σ is rational.

Intuitionists insist that the only evidence for such a sentence must consist in calculating two integers k and l such that σ equals k/l. But one cannot obtain this evidence from the mere knowledge that a contradiction follows from the assumption that σ is not rational. Also, we do not have that evidence and may never have it. Thus, intuitionists conclude, the rule of double negation elimination is not universally valid since otherwise we could deduce what k and l are from the information contained in the evidence for the double negated statement. Consequently, due to the rejection of this rule of inference, a few laws of classical logic stop being universally valid, most notably the principle of the excluded middle.

This argument appeals to subtlety. It is insisted that we should distinguish, in general, between evidence for the assertion of $\neg\neg p$ and evidence for the assertion of p. To uphold this difference it is demanded that the rules of inference should preserve the 'availability of evidence' instead of preserving verification-transcendent truth. But could this appeal to subtlety convince the intuitionist's opponent? While confronting the dilemma: subtlety or bivalent truth, the opponent will inevitably take the latter horn, saying that the objections to the bivalent truth that it blurs the difference between sorts of mathematical evidence are not strong enough. Hence we need to investigate what reasons Brouwer might have for repudiating bivalent truth.

3.8.3. *Indeterminacy or Infinity*

It is natural to think that Brouwer's rejection of the concept of verification-transcendent truth was motivated by his repudiation of realism. Just assume, as Brouwer did, that any facts, mathematical or otherwise, are constructions of the knowing subject carried out in accordance with some rules. So, you may say, there are no 'non-constructed' facts. If you now assume that facts are expressible by sentences, you could be inclined to repeat after Brouwer that "there are no non-experienced truths" and explain this slogan by saying that any sentence for which the corresponding construction does not obtain is not classified as true. Given this inclination, let us look in detail at whether such ontological constructivism may be seen as providing grounds for rejecting bivalent truth. One way of approaching this is to suppose that future constructions are not yet determined in the present state of knowledge. Yet, the hard thing is to grasp what the indeterminacy of mathematical constructions could mean. Mentioning indeterminacy makes one think of another revision of logic, namely Łukasiewicz's rejection of bivalence and introduction of a third logical value. As he reports in 'On Determinism' (1961),[36] it was a belief in the indeterminate course of future events that made him give up bivalence and postulate a third logical value. Thus, one might hope that Łukasiewicz's motivations could shed light on what mathematical indeterminacy can be. As the sentences he was concerned with express what happens, obtains or takes place at a time T, we may for convenience denote these sentences by letters with subscripts indicating the times to which they relate. What Łukasiewicz opposes is the following thesis:

(1)   If p_T, then it is true at any time t prior to T that p_T.

To illustrate this, if it rains in Cracow on Nov. 18, 1999, then it is true at any earlier time that it rains in Cracow on Nov. 18, 1999. Łukasiewicz calls the above claim the determinism thesis, though perhaps its more adequate name is fatalism. The thesis may be derived from two premises, of which the first is a temporal version of the principle of the excluded middle,

(2)   It is true at t that p_T or it is true at t that non-p_T,

while the other is an intuitive claim:

(3)   If it is true at a time t prior to T that p_T, then p_T,

which is intended to express the conviction that facts that obtained at some time, do not change later.

Now, what gives a sting to the determinism thesis is Łukasiewicz's analysis of the expression 'it is true at t that ... ', which he construes as meaning: 'at t a fact obtains which causes ... to be the case'. In effect the innocent looking determinism thesis turns into a substantial claim:

(4)   If p_T, then at any time t prior to T, a fact obtains that causes p_T to be the case.

This full-fledged determinism, however, appears incorrect to Łukasiewicz, as he disbelieves that any fact has its cause at any past time and also affirms that humans are essentially free to affect the future course of events. Consequently, he rejects it, which, given his analysis of 'it is true at t that ... ', makes him renounce bivalence and introduce a third logical value for indeterminate sentences, that is, sentences about the future.

Trying to find a path from Brouwer's ontological constructivism to his repudiation of bivalent truth we may thus ask whether, despite substantial differences between Łukasiewicz's three-valued logic and intuitionistic (two-valued) logic, there is something in Brouwer's philosophy that may play a role analogous to Łukasiewicz's indeterminate future events or indeterminate sentences about the future. Or, to ask this question in other terms, what motivates Brouwer's rejection of (1), i.e.,

If p_T, than it is true at any time t prior to T, that p_T?

That Brouwer must reject it is clear from his doctrine of 'no non-experienced truths'. Just think of a statement that was first proved on, say, Jan 1, 1996—for Brouwer it was not true at any time prior to this date. In the case of Łukasiewicz, his rejection of (1) was a consequence of the belief in the indeterminacy of future events and his rather special construal of the phrase 'it is true at t that'. Perhaps the first answer that comes to mind is that among mathematical problems for which at present we do not know the solutions, there are also truly undecidable statements of which it *cannot*, in some absolute sense, be known whether or not they hold. Accordingly, a truly undecidable mathematical sentence would be akin to a sentence about a future event, for which at a given time there is nothing that causes it to happen, or prohibits its happening. Importantly, on this construal we have a positive declaration of indeterminacy: future events are not determined by present ones, some yet undecided mathematical problems are not determined as they are truly undecidable. And, since truth is closely related in intuitionism to knowledge of proof, while falsity is related to knowledge of refutation, sentences expressing truly unsolvable statements would be akin to Łukasiewicz's indeterminate sentences (that is, those possessing a third logical value). This proposal, however, is hardly tenable on intuitionistic standards since the intuitionist cannot meaningfully claim that we have truly unsolvable statements, if by this he understands that neither a construction proving the statement, nor a construction proving its negation can be carried out. Thus, Brouwer's repudiation of verification-transcendent truth cannot be understood as following from the discussed sort of indeterminacy. One may try to improve on the approach that invokes indeterminacy by emphasizing another conceivable similarity between Brouwer's and Łukasiewicz's doctrines. As sentences about the future that are indeterminate at some time become in due time either true or false, so statements of intuitionistic mathematics that are not assertible in a state of information may later become either true or false. This analogy is, however, only superficial. Łukasiewicz's opposition to bivalence stems from envisaging states of affairs of which at the present time physical reality does not determine whether or not they will obtain. Typically you may think of a quantum particle before it takes one of two possible trajectories, say, going through one or the other slit. Now, if we construe analogously the intuitionistic objection to bivalence, we need to say that a mathematical statement does not obey bivalence because somehow it is not yet decided in the mathematical realm which way this statement will go: will it be true, or false? This supposition, however, appears groundless, if not absurd.

The failure to account for intuitionistic objections to bivalence by invoking some sort of indeterminacy makes us suspect that while reconstructing the rationale for dispensing with bivalence, we may have confused indeterminacy with intuitionistic reservations about actual infinity. Indeed, Brouwer's protests against the validity of classical logic always take place against the background assumption that the domain involved is infinite. Similarly, his rejection of merely linguistic methods of introducing mathematical objects is restricted to the use of them in investigations concerning infinite totalities. That is, as long as a domain is finite, by announcing a predicate, say ψ, we divide it into those objects that satisfy ψ and those that do not, so that for any given object t from the domain either $\psi(t)$ or not-$\psi(t)$ holds. The proponent of only potential infinity disagrees that this method can be applied to infinite totalities and argues that in doing so one is treating infinite totalities as complete. From his perspective, however, their incomplete character is their essential feature. It is an essential feature of the natural numbers that they are never finished, never complete, and it is not a limitation of our capacities that we cannot survey them all, but a reflection of the very way they are. Metaphorically speaking, even the arithmetical knowledge of an omniscient God cannot be based on His seeing or surveying all the natural numbers; if this Being knows some arithmetical theorem, He derives this knowledge similarly to the way we do so, that is by induction. Given that one thinks of the natural numbers in a way that rules out regarding them as a complete totality, one needs to adjust the concept of truth appropriately, or explain what it means that an arithmetical statement holds. Indeed, the idea of a statement being true in such a way that its truth might be not known in any state of knowledge is a bit untrustworthy, as it tends to combine with a picture of a complete arithmetical reality that makes the statement true, no matter whether it is known or not. For someone who rejects this picture of mathematical reality and accordingly opts for linking truth with provability, there appear to be two options to characterize what it means that a mathematical statement holds. Beyond the first option there lies the observation that if we relate truth to actual provability, some statements that have not been proved for only contingent reasons will not be classified as true and similarly those that have not been proved to lead to a contradiction will not be considered false. Thus, we form an image of an idealized mathematician, or God equipped with intuitionistic abilities to prove theorems but not capable of surveying the whole infinite mathematical realm. That is, we imagine that this Being is incomparably faster at proving theorems than we are, and perhaps that He even knows every proof and disproof in advance; nevertheless there is no reason to believe that for every mathematical statement He has either its proof or a proof of its negation. It is then said that the statement, say ψ, is true if that Being knows its proof, while ψ is false if He knows a proof of not-ψ. As we do not have any reason to believe that every mathematical problem is solvable, even by such a Being, we consequently refuse to accept that that any mathematical statement is either true or false. If the classical mathematician takes the bait of characterizing truth in terms of such idealized solvability, then the discussion about bivalence boils down to a debate about whether any mathematical problem is solvable, with the intuitionist expressing his reservations. It is nevertheless highly debatable whether the reservation is strong enough to impinge on bivalence. We may imagine both sides debating over an intuitionistic weak counterexample, and at some point the classical mathematician asking his opponent, 'So, you think that this statement, which is not decided at present, can neither be proved nor disproved by your omniscient Being?' And, as the intuitionist cannot answer it in the affirmative, since strong undecidability is for him either meaningless or absurd, he has to limit himself to expressing ignorance along the lines of, 'I do not know if the problem is solvable; I do not have grounds for believing that any problem is solvable'.

The other option is to leave the concept of mathematical truth to the classical mathematician, as it tends to bring forth the issue of distinguishing between a statement not being true for mathematical reasons and not being true for contingent reasons. To the former option, it will be objected that it still does not sufficiently mirror the constructive character of mathematics, as it smuggles in the image of a predeterminate three-fold division of mathematical statements into those that can be proved, those whose negation can be proved, and those for which none of the above obtains. The present option recommends considering mathematical objects as constructed objects, or rather as emerging indefinitely from some processes of construction. Now, the supposition that the construction is free appears to be incompatible with the idea that the object has a property, although an already generated finite segment of that object does not allow us to prove that the property holds. Thus, instead of seeking a constructive concept of truth, the proponent of this view decides to explicate what it is that the statement holds in terms of its being assertible in a given state of knowledge. Roughly, the statement holds if it can be asserted on the grounds of knowledge about the generation of the already achieved segment of the relevant infinitely proceeding object. There is a slight positivistic flavor here, as the proponent can be understood to be saying that we do not know which picture of the natural numbers is correct—is it an independently existing complete totality, or an incomplete but predeterminate one, or a free and non-predeterminate one?—so we had better dispense with the notion of truth and content ourselves with saying that a statement can be *asserted* if the corresponding construction is effected and explain further what sort of construction is required for the assertion of statements of various logical forms. Somewhat similarly, an intuitionist may say that he studies mathematical objects as given, while accusing classical mathematicians of assuming the Platonic existence of mathematical objects. This line of argumentation has been elaborated by Brouwer's student, Arend Heyting, and for this reason we will investigate it in more detail when we come to his arguments for intuitionism. Here let us only comment on a problem inherent in this view. To be attracted by intuitionistic revision, we need to resist the inclination to conceive of properties or relations of the natural numbers as already obtaining or not, given the way the natural numbers are introduced. For instance, if the way the natural numbers are introduced is correctly encapsulated in Dedekind's axioms, there is the tendency to think that they decide whether or not, say, Goldbach's conjecture holds, no matter whether we know it or not. To resist this tendency, it is helpful to consider freely generated choice sequences, as it is hard to conceive that such a freely progressing sequence has a property if this cannot be known on the ground of the available initial segment of it and the restrictions already imposed on its generation. Consequently, the view that mathematical statements that pertain to choice sequences may hold, though at present we cannot know this, is given up as inadequate for the choice sequences. Instead it is said that the statement holds if we can know, on the grounds of information available at a given stage of knowledge, how to prove it.

We have here a relatively strong argument against bivalent truth, but unfortunately one that relies on objects, like choice sequences, that are rather alien to classical mathematics. And, it is weak counterexamples that should persuade a potential convert to intuitionism to repudiate bivalence. On their own, however, the counterexamples have rather limited persuasive power. The classical mathematician can obviously argue that they do not offer any grounds against holding that each statement is either true or false, as they only show that neither a statement nor its negation is assertible in a given state of knowledge. Thus, they convince only those who are already convinced, that is those who agree with the intuitionist on his objections to the legitimacy of the concept of bivalent truth when applied to statements concerning infinite and freely proceeding objects. However, for someone who is convinced that bivalent truth should be replaced by assertability, the intuitionist has more powerful arguments to the effect that some instances of the generalized principle of the excluded middle and some instances of the generalized law of double negation elimination are contradictory.

3.8.4. *Strong Counterexamples to the (Generalized) Excluded Middle*

In the intuitionistic theory of real numbers it is possible to prove much stronger results to the effect that from an instance of the universally quantified principles of the excluded middle or an instance of the generalized double negation elimination contradictions follow. Given the intuitionistic understanding of negation, this means that sentences of the form

$$\neg \forall_x (P(x) \vee \neg P(x)) \quad \text{or} \quad \neg \forall_x (\neg\neg P(x) \Rightarrow P(x)),$$

are intuitionistically provable. These results constitute the so-called strong counterexamples to the laws of classical logic. Let us discuss the theorem stating that a contradiction follows from the assumption that for any choice sequence of natural numbers it can be decided whether or not it consists of zeros only; in symbols:

$$\neg \forall_\alpha (\forall_k \alpha(k) = 0 \vee \neg \forall_k \alpha(k) = 0)$$

where the Greek letters $\alpha, \beta, \gamma, \ldots$ range over choice sequences of natural numbers, $k = 0, 1, 2, 3, \ldots$, symbol $\alpha(n)$ denotes the n-th element of α and α_n stands for the sequence $\alpha(0), \alpha(1), \ldots, \alpha(n-1)$. The exposition is based on (EI, pp. 79–85). Since the argument is rather technical, for the proof and further discussions of the theorem the interested reader is referred to Appendix.

The argument for the theorem appeals to a principle peculiar to intuitionism, known as the $\forall\exists$-continuity principle, its peculiarity stemming from the quantification over choice sequences. To come to grips with the principle, assume first that for any choice sequence α, a natural number n can be found such that it is in relation B to α; in symbols $\forall_\alpha \exists_n B(\alpha, n)$. Assume further that the relation B is extensional in respect to choice sequences, $\text{Ext}_\alpha B(\alpha, n)$, that is, if it can be proved both that a sequence β is extensionally identical to α and that $B(\alpha, n)$ holds, then it can be proved as well that $B(\beta, n)$ holds. Now, $\forall\exists$-continuity principle states that if for any choice sequence α a natural number n can be found such that $B(\alpha, n)$ holds and B is extensional in respect to choice sequences, then one may find a *function* e^* that ascribes natural numbers to choice sequences on the basis only of initial segments of these sequences, and such that $n = e^*(\alpha)$ and the relation $B(\alpha, e^*(\alpha))$ obtains. In symbols, we have

$$\forall_n \operatorname{Ext}_\alpha B(\alpha, n) \wedge \forall_\alpha \exists_n B(\alpha, n) \Rightarrow \exists_{e^*} \forall_\alpha [e^*(\alpha) \text{ is defined} \wedge B(\alpha, e^*(\alpha))].$$

Consider now a disjunction $C(\alpha) \vee D(\alpha)$. Clearly, if both disjuncts are extensional, so is the disjunction. Suppose now that $\forall_\alpha (C(\alpha) \vee D(\alpha))$ holds, which means that a procedure is known that decides for any α whether $C(\alpha)$ or $D(\alpha)$ holds. Thus, we may supplement the disjunction by demanding that if $D(\alpha)$ does not hold, then we put it that a number n is equal to 0, but if $D(\alpha)$ holds we put it n is 1. Since if $D(\alpha)$ does not hold, $C(\alpha)$ must hold, we obtain:

$$\forall_\alpha \exists_n ((C(\alpha) \wedge n = 0) \vee (D(\alpha) \wedge n = 1)).$$

We note thus that the following implication holds:

$$\operatorname{Ext}_\alpha C(\alpha) \wedge \operatorname{Ext}_\alpha D(\alpha) \wedge \forall_\alpha (C(\alpha) \vee D(\alpha))$$
$$\Rightarrow \forall_n \operatorname{Ext}_\alpha ((C(\alpha) \wedge n = 0) \vee (D(\alpha) \wedge n = 1))$$
$$\wedge \forall_\alpha \exists_n ((C(\alpha) \wedge n = 0) \vee (D(\alpha) \wedge n = 1)).$$

Accordingly, plugging the consequent of this implication into the premise of $\forall\exists$-continuity principle we obtain the continuity principle for disjunction:

$$\operatorname{Ext}_\alpha C(\alpha) \wedge \operatorname{Ext}_\alpha D(\alpha) \wedge \forall_\alpha (C(\alpha) \vee D(\alpha))$$
$$\Rightarrow \exists_{e^*} \forall_\alpha [e^*(\alpha) \text{ is defined} \wedge ((C(\alpha) \wedge e^*(\alpha) = 0) \vee (D(\alpha) \wedge e^*(\alpha) = 1))].$$

Next we plug in for $C(\alpha)$: $\forall_k \alpha(k) = 0$ and for $D(\alpha)$: $\neg \forall_k \alpha(k) = 0$. Since both these relations are extensional, we get:

$$\forall_\alpha (\forall_k \alpha(k) = 0 \vee \neg \forall_k \alpha(k) = 0) \Rightarrow \exists_{e^*} \forall_\alpha [e^*(\alpha) \text{ is defined}$$
$$\wedge [(\forall_k \alpha(k) = 0 \wedge e^*(\alpha) = 0) \vee (\neg \forall_k \alpha(k) = 0 \wedge e^*(\alpha) = 1)]].$$

Finally, it is shown that the consequent of the above implication leads to a contradiction. Hence, it is proved that it is impossible that

$$\forall_\alpha (\forall_k \alpha(k) = 0 \vee \neg \forall_k \alpha(k) = 0).$$

The crucial step in strong counterexamples to the principles of classical logic is the continuity principle, whose proof depends on a method peculiar to intuitionism, namely reflecting on the possibility of gaining information that a premise holds. We ask for instance how one may know that for any choice sequence α a natural number can be found such that it is in a relation B to α. Importantly, this is information about freely generated infinitely proceeding sequences that are never given in their entirety (with the exception, however, of law-like sequences). And the admissibility of such sequences motivates one to construe a statement about a property of choice sequences as concerning the possibility of gaining information that the sequences have that property. Such a motivation is lacking in the case of the classical mathematician who conceives of infinite sequences of natural numbers as extensional, completed objects. Thus, he readily accepts the possibility that a sequence may have a property, though we do not know, or even may be unable to know this. Now, if you think of intuitionistic essentially free choice sequences, the above picture turns out to be inappropriate. For, given that the generation of the choice sequence is essentially free, and that it is therefore not determined in advance whether or not it will have a given property, it is hard to give any meaning to the claim that a free sequence has the property, although it cannot be derived from our knowledge of an initial segment of the sequence and initial restrictions on its generation, that it has this property. This brings us to the second feature of the argument for the above strong counterexample, namely, the assumption of intuitionistically interpreted logical connectives and quantifiers. The assumption, however, is supported by assuming quantification over free choice sequences. If α ranges over free choice sequences, it is hard to read $\forall_\alpha \exists_n B(\alpha, n)$ in any other way than 'a method is known of finding, for any α a number n such that $B(\alpha, n)$ holds'. In a similar vein, it is hard to conceive that a disjunction $P(\alpha) \vee Q(\alpha)$ holds of a choice sequence α, while it is known neither that $P(\alpha)$ holds, nor that $Q(\alpha)$ holds. A similar inclination attaches to negated statements concerning choice sequences. If these intuitions are right, they support the intuitionistic view of the excluded middle. On the other hand, they appear to be counterbalanced since it is not clear why, for a given choice sequence, one should deny, on the grounds of its being freely created, that it either has a property or not, no matter how little we know about one or the other possibility. Thus, there are two ways of blocking the reasoning that leads to the strong-counterexamples. First, the opponent may declare outright that the reasoning assumes intuitionistic understanding of the logical constants, and for this reason, the counterexample is illusory. This is not a very cogent move, since the intuitionist will plausibly argue that the free character of choice sequences requires his interpretation of the logical connectives and abandonment of bivalence. Second, it may be objected that choice sequences are not legitimate mathematical objects since they are somehow 'human-dependent' or 'time-dependent'. This course of action goes deeper since what motivates the intuitionistic understanding of the general quantifier in the above argument is the assumption of choice sequences. The sequences may be called 'human-dependent' because in introducing them one invokes the idealized human subject or some other agent that is capable of freely generating numbers. On the other hand, at any stage of generation, the knowledge of a choice sequence changes, and since the generation is truly free, there is little sense in claiming that the whole sequence *is* in its entirety. That is the reason for calling choice sequences time-dependent. Their 'human-dependent' side can, however, be reduced, at least to some extent. Although Brouwer's papers do contain suggestions that choice sequences are made by an idealized human subject, with an implicit indication that it is the same agent that generates and mathematically investigates choice sequences, one may perhaps devise a computer program simulating the generation of choice sequences, or if you prefer, the creative function of the Brouwerian subject. Such a program, apart from generating a series of numbers that accord with previously introduced restrictions, should 'decide' at any stage whether or not a new restriction is to be imposed, and, if so, 'select' such a restriction from a set of some restrictions. The output of the program, that is, a sequence of ⟨number, restriction⟩ pairs, corresponds to what the subject may know about a choice sequence. Nevertheless, there is a weak point in the idea that a choice sequence may be identified with the output of some computer program and the activities of the Brouwerian creative subject simulated by a machine. The choices should be essentially free, this being most readily understood as the requirement that the numbers be chosen in a truly random fashion. This is, however, precisely what the computer cannot achieve as it generates pseudo-random numbers only.

We have also the option of opposing the concept of choice sequence on rather ideological grounds, claiming that in the mathematical realm time can play no role. Whether or not this dictum should be accepted is a matter of taste and philosophical preferences. Yet, a demonstration that choice sequences are not legitimate objects of mathematics would still leave the classical logician facing a nagging problem. Even if choice sequences are not mathematical objects, they are still objects of some sort. And it seems that a discourse about them violates the laws of classical logic. But is this really so? We have only the right to assert that if in certain instances of some laws of classical logic which concern choice sequences, the logical connectives and quantifiers are understood intuitionistically, then these statements are contradictory. Thus, the crucial question is whether the quantifiers must be so construed in statements concerning choice sequences. To this the intuitionist has the cogent answer that in the discourse concerning choice sequences we cannot think of statements as determinately either true or false, and for this reason the logical constants cannot have their classical meanings.

Let us briefly summarize this chapter. Brouwer's views that concern language, logic and the relation between logic and mathematics form a rather consistent doctrine, significant for his arguments for the revision of mathematics. One part of it is the refusal to admit that language has any other role than being a means of communication, which is understood as influencing the actions of others or invoking appropriate images in other minds; any idea that language is also a medium of the adequate representation of mathematical constructions or a medium of mathematical proofs meets Brouwer's disapproval. Consequently, he professes a sharp separation between mathematics and language and rejects 'linguistic' methods of practicing mathematics: the axiomatic-deductive method and, most strongly, Hilbert's Program. As to rules and laws and logic, he considers them to be idealized regularities obtaining in reports on the carrying out of mathematical constructions. This stance, given the assumption of what sort of mathematical constructions are described by sentences containing a given logical constant, makes him renounce the principles of the excluded middle and double negation elimination since some instances of them cannot be asserted, as the required constructions are not known. This argument is rather unconvincing, as it draws on a view of logic hardly acceptable to the classical mathematician. What is at stake in the controversy is the legitimacy of bivalent truth as applied to mathematical statements. According to one interpretation, one may try to derive the rejection of bivalence from the assumption that the unknown constructions are somehow indeterminate as to whether or not they can be effected; it turns out, however, that this supposition lands us in absurdities. What remains is to derive the rebuttal of bivalence from Brouwer's doctrine that infinity is only potential and his assumption of objects whose generation is essentially free, that is, choice sequences. This path appears to be promising, and we will take it up while examining the more elaborate version of this argument, which emerges from Heyting's writings. Now, if there is such a path, the assessment of Brouwer's project depends on the issue of the legitimacy of actual infinity and choice sequences. Clearly, these issues can hardly be solved in philosophical speculations; nevertheless one may hope to solve them by comparing how fruitful the competing schools of mathematics are, that is, those which admit actual infinity and those which admit only potential infinity but accept choice sequences. Here the notion of a mathematical theory being fruitful, though far from precise, relates both to some inner mathematical criteria and applicability of mathematics in natural sciences.

4. INTERSUBJECTIVITY IN BROUWER'S CONCEPTION OF MATHEMATICS

While asking whether Brouwer's philosophy allows for communication of mathematical knowledge, we should carefully distinguish two related issues. First, one may query whether mathematics as practiced by Brouwer is more likely to lead to *practical* difficulties in communication than classical mathematics. Secondly, one may pose a purely philosophical problem by arguing that if mathematics were in actuality practiced in accord with Brouwer's philosophy, mathematical results could not be communicated. In asking this question we put aside all facts from the history of intuitionism, concentrate on Brouwer's views on mathematics and related matters and ask whether his conception of mathematics does involve the consequence of making mathematical results incommunicable. Thus, the philosophical issue is to a large extent independent of the practical one: even if intuitionistic mathematicians seem to communicate quite well, a philosopher occupied with the philosophical problem may see this as only an illusion. Conversely, the philosopher may concede that mathematicians communicate but add that this is so only because, after all, their cognitive activities diverge considerably from cognitive activities as assumed in Brouwer's doctrine. We shall deal below with the philosophical problem and completely ignore the first, practical, issue. However, we need first to examine, and if possible dispel, some common accusations against Brouwer's doctrine, as the issue of their validity has a direct bearing on our investigation.

4.1. *Psychologism, Subjectivism, Solipsism?*

Rumor has it that Brouwer's conception of mathematics is prone to 'psychologism', that it is a 'subjectivist' doctrine or even that it advocates 'solipsism'. Although these 'bad names' have a familiar clang for philosophers, we shall start with a discussion of these positions. It will prove important for our investigations since, as it turns out, what is really at stake in these accusations is the question of intersubjectivity, that is the question of whether Brouwer's conception of mathematics allows for the unity of mathematics as one discipline practiced by many mathematicians whose results are communicated by written or oral means. Thus, in a somehow roundabout way, by discussing these objections to Brouwer's philosophy we shall return to the issue of intersubjectivity and test whether or not it can be satisfied in Brouwer's conception of mathematics.

The first thing to note is that psychologism is a species of naturalism.[37] According to this doctrine, logical and mathematical objects, proofs, theorems, and constructions are somehow in the mind; they are mental entities. One may express this in a more language-oriented way by saying that logical and mathematical statements are about mental phenomena. As such, psychologism has two components. The first is the ontological thesis about the mental status of logical and mathematical objects. The second is an epistemological claim that in order to justify mathematical or logical theorems one needs to examine one's mathematical or logical objects, constructions and proofs. Again, the latter claim may receive two different readings. It may be understood as a rather innocuous demand that evidence for those statements should be produced by reflecting upon the *contents* of appropriate mental acts and processes. This is a rather trivial and innocent thesis because, after all, what we actually do in mathematics or logic is reflecting, thinking or understanding. Much more controversially, the thesis can be conceived as a requirement that a part of the justification of a mathematical or logical theorem should consist in empirically tracing the psychological *origins* of 'mental concepts' or 'ideas' involved in that theorem. According to this demand, for instance, if a person sets about to prove, say, the Pythagorean theorem, he should investigate whether or not his mental concepts of triangle, right angle, and others are legitimate, and that question should be answered by investigating processes of the formation of those concepts. In proving this theorem one should then discover that the mental concept 'right triangle' comprises another concept, namely 'the sum of all angles being $180°$'. The thesis about the 'genealogical' character of justification can be traced back to Locke's and Hume's writings.

Contrary to what our example may suggest, psychologism was a popular philosophy of logic in the nineteenth century, but it rather did not concern mathematics. Traditionally understood, logic is more prone to being psychologistically interpreted than mathematics due to its popular characterization as the science investigating laws of reasoning. From this definition of logic it is not far to the slogan "laws of logic are laws of thought" which, given that thought is assumed to be a flesh-and-blood person's thought, commits one to a view that these laws govern specific mental processes, known as reasoning. In the nineteenth century psychologism, in this or some other version, kept a strong hold on the imagination of a variety of logicians and psychologists, some of whom set about to discover laws of logic by introspective studies of processes of thinking. The doctrine fell into disrepute at the turn of the century, as Frege, Husserl, Łukasiewicz and others launched an attack on its crucial tenet, that is, the claim that logic studies laws of thought, where thought is understood as a flesh-and-blood person's thought.

At this point, let us notice that on its own the ontological thesis of psychologism is rather philosophically innocuous, since without being supplemented by other tenets it does not lead to philosophically suspicious consequences. Also, as we have already remarked, the weak epistemological thesis of psychologism, that evidence for mathematical statements should be found by reflecting upon the contents of specific mental processes, is harmless and rather realistic. Thus, what gives a sting to psychologism is the genealogical thesis, as it endangers the requirement of intersubjectivity. After all, our acts of thinking, origins of concepts, and their formations may differ considerably. Consequently, to know whether Brouwer was prone to militant psychologism and to find whether accusations about the lack of intersubjectivity are right, one needs to know if he accepts the genealogical claim about the nature of justification.

It is clear that his way of writing invites, so to speak, the accusation of psychologism. First, the idea that mathematics deals with mental constructions of the subject is most readily understood as a claim about the ontological status of mathematical objects, namely, that they are 'mental'. Second, the requirement of a 'languageless' construction as the evidence for mathematical truths closely resembles the epistemological thesis of psychologism, that justification should rely on reflection upon mental processes. However, it is highly doubtful whether one may rightly attribute to Brouwer the 'genealogical' thesis that justification should be sought in studying the psychological origins of mental concepts or ideas and the processes of their formation. As far as we know, no claim of Brouwer's gives any hint of a reason for ascribing to him this belief. Quite to the contrary, one may find in rejected parts of his doctoral dissertation (van Stigt 1979, p. 397) a distinction between *mathematical system* and *emotional content* coupled with mathematical system. He maintains that the former is shared by many mathematicians, while the latter changes from person to person. A page later he phrases this distinction a bit differently, as he distinguishes between objective *relations* in a mathematical system and their subjective *representations*. Intuitionistic mathematics aims at investigating mathematical systems or their objective relations and not the accompanying emotional contents or representations of those relations. For that reason, studies of mental processes in their entirety are irrelevant from the intuitionist standpoint, even if they are somehow linked to practicing mathematics. What matters are relations obtaining in a mathematical system and these are believed to be identical, despite the idiosyncrasies of individual mathematicians. Thus, the identification of justification of mathematical statements with investigations of the actual formation of mental 'ideas' ('representations' in Brouwer's terms) is contrary to his teaching. Consequently, one should not ascribe genealogical psychologism to Brouwer. It should also be observed that Brouwer's distinction between the mathematical system and the emotional content (or between relations and their representations) is akin to traditional distinctions introduced at the turn of the century which allowed philosophers to discard the doctrine of psychologism. Thus, Frege draws a distinction between thought and thinking, Husserl between the act of conferring meaning and the ideal meaning, Łukasiewicz between the act and its object. Needless to say, these distinctions are now a commonplace in philosophy.

The second epithet used to describe Brouwer's doctrine, solipsism, seems to come from his discussion of the problem of other minds. To repeat, on our construal, Brouwer's main contention is that the knowing subject, who due to the causal attention organizes data into complexes and postulates objects and causal links, cannot ascribe mind to any of 'his' objects. As we have argued, however, this stance does not exclude the possibility of experiencing other minds differently, that is without the mediation of the causal attention. For this reason, one cannot honestly characterize Brouwer's position as solipsism, if that term is used with its ordinary meaning, as the rejection of the existence of other minds. But since everything in Brouwer's philosophy, apart from the knowing subject, its sensations and its various capacities, is constructions and postulated objects, the term 'solipsism' may mistakenly have been used to emphasize the idealistic character of that doctrine. The words 'subjectivism' or 'subjective idealism' characterize this doctrine much better, however.

Subjectivism is usually explained as an epistemological position that assumes the results of cognitive operations to be dependent on or determined by particular ways of gaining knowledge. Depending on the sort of subjectivism, results of cognitive processes are believed to be determined by idiosyncrasies of individual sensory apparatus, by emotional states of the subject, by social or historical circumstances, by the biological setup of the subject and so on. In order to be lured into this view we need to entertain seriously the possibility that if a person had a different sensory apparatus, or if his mind were constructed differently, or if he lived in some other social circumstances, then his body of knowledge would differ considerably from what he actually knows. Moreover, for that hypothetical person, other truths might hold than for the actual one.

One motivation for calling Brouwer's philosophy 'subjectivism' is presumably his idea that the subject who uses various 'capacities' is the creator of so-called material objects, mathematical constructions, and causal chains. The question whether anything in external reality corresponds to a construction, (e.g., to a causal link or object) receives a negative answer: causal links as well as objects exist only in the mind. On the other hand, the supposition that something may exist in external reality that cannot be given to the subject by means of its causal attention is rejected. However, in order to find out if the motivation is not misplaced, we need to ask if, and if so, in what sense, knowledge is assumed in Brouwer's philosophy to be relative in respect to the cognitive apparatus of the subject. Brouwer indeed emphasizes that the intuition of two-ity is the sole generator of mathematical objects, while the causal attention is the only means of acquiring knowledge about other matters. Thus, in order to pose the problem, we need to introduce the rather meaningless supposition of a hypothetical subject whose intuition of two-ity and causal attention operate somewhat 'differently' than the two abilities of the Brouwerian subject. In Brouwer's doctrine, a person knows that a mathematical theorem holds since he or she has created an appropriate construction in accordance with the intuition of two-ity. Were the intuition of two-ity different, that person would not be capable of effecting the construction mentioned and, accordingly, the mathematical theorem would not hold for him. Perhaps in this case the theorem would not be provable. We may illustrate it, with exaggeration, by saying that $2+2=4$ holds for a person who has the intuition of two-ity, but it is not certain whether it holds for somebody with a different intuition. Similarly, a person knows that a fact obtains, since in the causal attention various sensations have been organized and linked in a way that assures the subject that this fact obtains. But again, if the causal attention or the machinery of postulating and objectifying worked differently, some other fact would be constructed instead of the original one. However, as to the suppositions that a knowing subject might have an intuition different than the intuition of two-ity or that his causal attention might work in some other way, a strong case can be put that these suppositions are utterly meaningless in Brouwer's doctrine. Since the intuition of two-ity is a defining property of the mind, any mind is capable of repeating the same mathematical constructions, and accordingly, can agree with others about the assessment of mathematical constructions. Consequently, any two people have the same mathematical intuition as long as they have (or are) minds. The objection that somebody's mathematical intuition could stop producing new units is dealt with by claiming that such a human being would have neither a Brouwerian mind nor the intuition of two-ity. In the same vein, somebody who does not perceive a flow of sensations (because he is asleep or has taken drugs) is said to have no mind. Similarly, if two mathematicians obtain different and irreconcilable answers to the same mathematical problem, then assuming the difference is not attributable to a flaw of a memory, it is claimed that they result from a 'gap' in the functioning of the intuition, which amounts to saying that one or both the mathematicians have ceased to have minds. It is perhaps worth noting that the intuition of two-ity, 'the empty form of all change' plays the role of the Kantian pure form of the sensuous intuition, since it is responsible for making everything be presented to the mind as ordered in time. This is one reason why Brouwer used a more Kantian term in his earlier papers, speaking of the *intuition of time*. The other 'pure form', space, is already rejected in 'On the Foundations of Mathematics' (1907, p. 70). On the other hand, one may see in the capacities that Brouwer ascribes to the subject a reminiscence of Kant's *categories* although the structure of the subject's abilities, so meticulously described in *The Critique*, is scantily elaborated in Brouwer's philosophy. This perhaps mirrors his conviction that human capacities cannot be subjected to any 'mathematical system' (*ibid.*, p.71).

The reply is not only highly speculative but also invokes the nagging problem of the relation between the flesh-and-blood mathematician and the Brouwerian subject or mind. More specifically, one may ask how concrete abilities of a real mathematician are related to the intuition of two-ity and other capacities of the subject as postulated by Brouwer. These questions are reminiscent of the notorious problem of Kantian idealism, namely the link between transcendental ego and the psychological subject, and are equally difficult to answer. Nevertheless, if an intimate link between mind and intuition is really assumed in Brouwer's philosophy, his doctrine can hardly be characterized as subjectivism. Although knowledge is assumed to be relative with regard to the cognitive abilities of the subject, it is not a psychological subject but rather the 'transcendental mind'. And the concept of the transcendental mind is so devised as to guarantee that all subjects equipped with 'the mind' may arrive at the same mathematical constructions. To accuse Brouwer of subjectivism is like arguing that Kant's doctrine is a form of subjectivism because a person might have a different geometry, due to a different form of perception.

To sum up these investigations into allegations raised against Brouwer, it turns out that the most serious accusations are ill-founded. Brouwer accepts the ontological thesis of psychologism as well as the weak variant of its epistemological version. However, what gives psychologism its sting, that is the 'genealogical' claim about the justification of knowledge, is denied by Brouwer, as his distinctions clearly suggest. A very similar situation is found in the accusation of subjectivism. Again, Brouwer's doctrine seems at first glance to be a form of subjectivism, as he teaches that almost everything—mathematical constructions, objects, facts, and causal connections—is created by the subject. This means that in principle knowledge is assumed to be relative with respect to the subject, since a fact may obtain for one subject, but not for another. However, since the *subject* does not coincide in Brouwer's philosophy with a psychological subject, or a flesh-and-blood mathematician, and should rather be interpreted as Kantian *transcendental ego*, the alleged subjectivism or relativism disappears. Finally, the accusation of solipsism is simply mistaken as it is based on misrepresenting Brouwer's 'no plurality of minds' argument.

4.2. *Brouwer and Intersubjectivity: the Mentalist Condition*

Now we are ready to ask whether the mentalist condition of the intersubjectivity of knowledge is satisfied in Brouwer's conception of mathematics. To repeat, this condition requires human subjects to have in their perceptual contents, thoughts and memories, some elements which remain essentially the same in spite of individual differences among people and the changes they go through. In Brouwer's philosophy there are two sorts of resources that permit the condition to be satisfied. First, he makes a distinction between relations in mathematical systems and their subjective representations and argues that the former are essentially the same, despite being constructed by various mathematicians. Second, since in his doctrine the mathematical intuition is an inherent ability of the mind, all subjects equipped with a mind share the same mechanism for constructing mathematical objects. This means that for any mathematical object properly constructed, the construction can be repeated by any other subject provided only that it has a *mind*. Both the answers are closely related so as to form a consistent reply.

One may, however, object that Brouwer's distinction between (objective) relations in the mathematical system and their (subjective) representations is not justified by or founded on other fragments of his doctrine. Similarly, a scrupulous objector may find that the distinction, after having been introduced in a short passage of his doctoral dissertation (ironically, in a part that his supervisor rejected), was never repeated in his later writings. This may be taken to suggest that it was made *ad hoc* and later discarded. One possible line of defense is to point out that other similar distinctions, like those between the conscious act and its content, thinking and thought, or the act and its object are no better justified by or founded on the teaching of their proponents. Those distinctions were introduced either as an evident fact, or as a method of combating psychologism, and little effort was made either to justify them or to ground them in any elaborated doctrines. What is more, Brouwer's philosophy is distinguished by the fact that it offers an explanation of why mathematical systems and their relations, despite being constructions of the subject, are objective in the sense that all *minds* have the potential to repeat them. These are creations built in accordance with the intuition of two-ity; this intuition is responsible for the properties of mathematical systems. Other mental objects, like 'emotional contents' or subjective representations, are not traceable to its activity. Thus, not only does Brouwer's philosophy make the distinction necessary for the satisfaction of the demand of intersubjectivity, but additionally it attempts an explanation of why mathematical constructions are repeatable.

4.3. *How Can One Communicate about Mental Constructions?*

We have seen that facts, objects, causal links, and mathematical objects are creations of the subject in Brouwer's philosophy. Moreover, all subjects equipped with the Brouwerian mind have the ability to arrive at the same inventory of objects, or at least at the same inventory of mathematical constructions. This ensures that the mentalist condition of communication is satisfied; nevertheless, one may still be troubled with a query about how two mathematicians who have constructed the same mathematical object can communicate it in *language*. Clearly, this issue concerns the functioning of language, or more specifically, it poses the problem of how Brouwer's mentalist metaphysics can be supplemented by an account of language that would make the linguistic communication of mathematical results possible.

The worry about the intersubjectivity of mathematical results, given that these are mental constructions, may have two separate grounds. It may happen that Brouwer's conception of mathematics, his ontological and epistemological ideas, really *do* lead to the conclusion that mathematical results cannot be communicated. But it may equally well happen, and it is our contention that this is so, that the worry stems from some misconceptions or naiveté about what necessary conditions for successful linguistic communication really are. It appears that the worry about the intersubjectivity in Brouwer's conception of mathematics is motivated by a view of linguistic communication that contains the following three tenets. Firstly, this view assumes that people are capable of forming the same mental objects. Secondly, it holds that it is essential for communication that the person associates words with mental objects; as a result of effecting this association words start *referring* to mental objects. Thirdly, to account for words and their usage being taught to a person by the linguistic community, it is maintained that there are some regular and observable manifestations of someone's being in a given mental state, as a scream is a natural manifestation of someone's being in pain. If such manifestations are missing, however, one is left with an arbitrary decision by a person to name a discerned mental object with any word he likes. The arbitrariness may be mitigated cosmetically by arguing that a trainee reacts to other people's use of a word as taught. Nevertheless, as the community has no access to the trainee's inventory of mental objects, it is the trainee who has a final say about what the word taught is going to refer to.

The above picture may be faulted on each of the three counts. Let us, however, reflect first on the third one. On the supposition that words refer to mental objects, we are inevitably drawn into the puzzle of how denotations are established. And, if there are no regular manifestations of someone's having a given mental object, then we are apparently left with an appeal to private ostensive definition. But clearly, there is no such thing as a primitive and natural manifestation of having accomplished a mathematical construction. There is no parallel between pain and specific facial expressions or crying and, on the other hand, mental mathematical construction and its external manifestations. As a natural link between arriving at a mathematical result and a natural manifestation of this very fact is missing, it is tempting to say first that basic mathematical concepts and vocabulary are taught to a person in social circumstances involving ordinary physical objects and then add that the expressions so learned are, somehow mysteriously, made to refer to pure mental constructions, an underlying conviction being that it is somehow easier to learn to use words to talk about observable objects that can be pointed at than to master expressions that apparently stand for mental objects.

Let us reflect for a moment on how one could explain the child's mastering of mathematical expressions and concepts. A social account will presumably assume that in early stages of language-learning one masters expressions like 'a doggy', as well as 'two doggies', or 'three doggies', there being no essential difference between the abilities needed to master the former and the latter sorts of expressions. Further, the account will stress that part of basic mathematical training is purely mnemotechnical: the child is taught to recite a litany of numerals in an appropriate order, or to answer which numeral immediately proceeds a given one. Later, he is introduced to some basic arithmetical operations, like adding or multiplying, and is made to memorize the multiplication table. This rather mnemotechnical training goes hand in hand with learning to apply this knowledge to a practical situation, say, of counting the fruit in a couple of baskets, or calculating the number of legs on several pets. The story may be developed further by arguing that by acquiring abilities to evaluate distances or volumes a person comes to terms with the concept of rational number. Finally, as to the learning of more elaborate mathematical concepts and expressions, this account will attempt to display what actually happens between a teacher and the students in classroom situations.

Now, as the second assumption makes clear, no matter how good the social account is, it faces an apparently insoluble problem of accounting for linguistic communication. The assumption that it is essential for successful communication about mental objects that the expressions refer to them leads to the query on how expressions of socially acquired language are made to refer to one's mental objects. Given that a social account explains how we acquire basic mathematical concepts, say of small natural numbers, and accordingly, how we master the appropriate vocabulary (number words), it remains a mystery how words, say, number words taught in circumstances involving commonly observable objects like dogs, apples and flowers, receive a new role of *referring* to the pure mental constructions of the subject. This enigma is somewhat similar to that of a person communicating that he is in pain. Clearly, the child learns to correctly use the word 'pain' in interactions with others; it may be said that he thus learns a new pain behavior. Now, the question is whether one should go further and say that not only is a new pain behavior learned, but also the child decides what the word 'pain' *refers* to. And once we answer 'yes' to this question, the problem emerges of how the referring link is established. It appears thus that the main ingredient in the confused picture of linguistic communication we have been analyzing is the idea of expressions referring to mental constructions. Following Wittgenstein one may suspect, however, that if we have a social account of mastering mathematical expressions, the whole issue of how words refer to mental objects turns out to be superfluous.

Although our issue of establishing the word's link to the mathematical construction is reminiscent of some strands of Wittgenstein's private language argument, there is a significant divergence. On the view Wittgenstein attacks, words refer to what can *only* be known by the subject, whereas to the contrary, Brouwerian mathematical constructions can be formed by any qualified person. The similarity is nevertheless revealed in an idea, inherent to this picture of language, that after a mathematical expression (say, a number word) of the common language has been taught to a person to be used in the communication about observable objects, and his knowledge has been further extended in math classes by learning some mathematical theorems, this person confronts a mysterious but apparently extremely important decision as to which of his mental objects should be named by a given number word of common language. As a result of this process of naming, words acquire a double character: apart from belonging to common language by virtue of their being used in social circumstances, they additionally refer to someone's mental objects. Also, the decision of naming is supposedly of the utmost importance for any later communication. It is imagined that whenever the person decides to communicate a construction that he has effected, he first surveys its elements, then consults the memory as to what words are names of these elements, and finally utters a combination of the appropriate words. Since the words are (partly) words of the common language, his hearer can form in his mind the same mathematical construction as the speaker's by going through this procedure in reverse. Thus, it seems that the main objective of naming is correctly remembering what was named, as this seems necessary for the communication to succeed. However, the suspicion is that the whole issue of correctly recollecting an association between a name and a mental object is redundant for explaining the functioning of language and, in particular, linguistic communication. Suppose that as a result of a process of naming a person has made the word 'two' of a common language refer to a pure construction of his own. Now, imagine that he wants to communicate his newly effected construction and, while surveying it, he finds an element that, as he recollects it, bears the name 'two'. Can he be mistaken in recollecting its name? The temptation is to say yes. Nevertheless, as long as he correctly uses expressions of common language, which may involve in that case abilities to correctly estimate numbers of observable objects and correctly solve arithmetical tasks in math classes, the issue of correctly remembering the name of a given mental construction is idle. As long as he correctly uses the word 'two' in social circumstances, it does not matter whether he thinks the name should refer to this or that mental construction.

We need, however, to look more closely at our last supposition, namely, that a person may use language in a way that agrees with the established practice while still being mistaken as to his earlier decision about naming a given mental object. The case here envisioned diverges from Wittgenstein's examples in that it does not make clear what this naming, or rather carrying over of a word from common language to signify a mental object, consists in. On the one hand we may explain the process of naming along the lines familiar from Wittgenstein's *Investigations*, that is by arguing that in essence it is private naming, after all. In this approach, the fact that a mathematical expression is used in a community in a given way gives the trainee only an impulse to decide to take that expression for the name of some specific mental object of his own. In other words, this person's use of the word is not held responsible to the established practice of using this word in the community. It is as if a person who found out how the word 'pain' functions in the community, said nevertheless to himself: 'this word will refer to such and such sensations of my own'. And later, if it turns out that his use of the word 'pain' conflicts with its usage by members of the community, he balks and says: 'I'm talking of MY sensations, and only I know what name they bear.' In this case, despite an appeal to common language, we are clearly operating with private naming. On the other hand, one may imagine that although mathematical expressions refer to mental objects of a given person, he holds this role of the word to be responsible to the use of this word in the community. To take an example, suppose that a person has made the word 'two' of common language refer to his mental construction, but subsequently found out that his usage of 'two' in social contexts considerably diverged from that of other members of the community, as he was coming up with different estimations of numbers of visible objects, or presenting different results of calculations demanded in math class. Observing this discrepancy, he realizes that he uses this word incorrectly and decides to improve on his naming of mental constructions as well. This scenario exploits a similarity to a person who mistakenly decided to make the word 'pain' refer to headaches only, but nevertheless, as a result of noticing a discrepancy between his and others' uses of the word, decides to classify his other pains under this word as well.

However, no matter what picture of naming is assumed, whether the camouflaged private naming or socially corrigible naming, the problem of naming and correctly remembering which mental object has been so named is otiose for a model of linguistic communication. In the first scenario, as a person's naming is assumed not to be responsible to the established use of words in his linguistic community, there are no criteria of correctly remembering a mental object associated with a given name. If it appears to a person that he correctly remembers what object he earlier associated with a given word, then he correctly remembers what object he earlier associated with that word. Or, to put it in Wittgenstein's words, whatever seems right to this person, is right (*Philosophical Investigations* I, §258). To take another example of his, checking correctness of the association between a name and an object by merely consulting the memory of the process of private naming is like checking whether one correctly remembers the time of the departure of a train by forming an image of a page of a time-table, an image that cannot be independently tested for correctness (*ibid*, §265).

In this respect, the second scenario of naming fares somewhat better, as it permits us to introduce a distinction between the correct and incorrect bearer of a given name. Simply, a mental construction by a person is a correct bearer of a given name if that person's use of this word is in accord with its use in the linguistic community; otherwise the object attached to the word is an incorrect bearer. Nevertheless, the reason why people communicate is not that they correctly remember links between mental objects and their names. Whenever a person's correct remembering of such links is called in question, in order to resolve the problem one has to turn to that person's use of the expressions and find out whether his use is in sufficient agreement with the use established within the community. Thus, it seems that in this approach we travel along a vicious circle: in order to explain why people communicate about mental constructions, we postulated that they correctly remember what constructions bear given names, and, in turn, to explicate what correctness means in this context we appealed to the agreement in the usage of these words. Thus, the appeal to naming of mental constructions and correctly recollecting the effected associations was only a redundant detour. A part of an account of what it is that people communicate shall make it clear that they agree in their uses of language. Perhaps there is something more to linguistic communication; nevertheless, it is not to be identified with correctly remembering associations between words and mental objects, if this concept of correctness is parasitic upon agreement in use.

Till now we have been concerned with disclosing the tenets of a confused picture of language that suggest that the mere assumption of the mental character of mathematical objects prohibits people from communicating about them. We have identified the mistake with the idea that an essential part of the linguistic communication is the requirement of mental objects being named, which in turn follows from the demand that in order for successful communication to take place, words must refer. As it turns out, however, an appeal to reference to mental objects is otiose in explaining why people communicate by means of language. Does this mean that the appeal to the mental character of mathematical constructions is superfluous in Brouwer's conception of mathematics as well? One should distinguish between the role mental objects play in learning a language and their role in communication. Without various mental abilities a person presumably cannot form mathematical concepts, and accordingly, cannot learn mathematical expressions so as to be capable of performing calculations, solving mathematical tasks, proving mathematical theorems and drawing consequences from mathematical statements. Whether these abilities coincide with those assumed in Brouwer's doctrine is another matter. But if we pose the problem of linguistic communication, the whole issue of reference to elements of a mental inventory turns out to be irrelevant. This is somewhat parallel to the functioning of color-words. In order for a person to master color-words, he must have the ability to distinguish colors. On the other hand, in addressing the issue of linguistic communication, we need assume neither that color-words *refer* to sensations of colors, nor (even less) that sensations to which a given color-word purportedly refers should be similar in different people. Similarly, as to Brouwer's idea that mathematical objects are mental constructions, this has no bearing on the possibility of communication. Moreover, it may convincingly be argued that one should not attach a special importance to this claim of Brouwer's. After all, everything in his philosophy with the sole exception of the subject is a creation of the mind. As they do not belong to the external world independent of cognitive activities of the subject, these creations may with some justification be called *mental*. What deserves more emphasis than this appeal to mentalism is the doctrine of the 'languageless' character of mathematical constructions with its stress on the intuitive character of mathematics mirroring Brouwer's opposition to the formalistic vision of mathematics and, more generally, to securing mathematical certainty by means of logical laws and rules.

So far we have argued that viewing mathematical objects as mental constructions does not on its own prohibit the communicability of mathematical results. Accordingly, one cannot rightfully accuse Brouwer's philosophy of mathematics of making the learning of mathematical expressions impossible. There are no reasons to think that his conception of *mathematics* violates what we have called the Wittgensteinian condition of intersubjectivity. But would it really matter if it turned out that he explicitly declared communication of mathematical results to be impossible? As these views are extraneous to his conception of mathematics, one should rather consider them a puzzling peculiarity of a great mathematician that has no bearing on the tenability of the proposed revision of mathematics. We might improve on his conception by keeping to his idea of mathematics as a free mental construction, while rejecting the confused claims concerning the functioning of language. As a matter of fact, there was in Brouwer a trait of character that made him skeptical, for most of his career, about both understanding and cooperation between humans. Our contention is, however, that this stems rather from his personality, than from the conception of mathematics he advocated.

We may nevertheless take a closer look at Brouwer's remarks pertaining to language and communication, to see whether they call into question the intersubjectivity of mathematics. Although Brouwer acknowledged two roles of language, as a support for memory and a medium of communication, we shall concentrate only on the latter feature. An initial reason to be optimistic about the possibility of interpreting Brouwer's views so that they do not prohibit linguistic communication can be found in his vision of how mathematicians communicate. In his doctoral dissertation he states:

People try by means of sound or symbols to originate in other people copies of mathematical constructions and reasonings which they have made themselves (...). (1907, p. 73)

This idea is also repeated in his later papers. Incidentally, the concept of evoking constructions matches well a basic category accepted in the Significs movement and used to discern various levels of language. In 'Signifische Sprachforschung' one learns that

(...) [T]he so called meaning of words is exclusively determined by their bearing on what the speaker expects, or on what the hearer experiences (...). (1919, p. 223, my translation)

Consequently, various levels of language are discerned in accord with how words affect speakers and hearers. It should be noted that in the light of Brouwer's early distinction between mathematical relations that may be shared by various mathematicians and their subjective representations, a successful process of communication leads to the production in the mind of a hearer of the same construction as that in the mind of the speaker. This model of communication, as Brouwer frequently stresses, does not exclude the possibility that errors and mistakes occur in communication; moreover, even an understanding of hypothetical subjects equipped with unlimited memories is endangered once they have recourse to language (1933, p. 443). Our question, however, is not to what extent linguistic communication is fallible, but rather whether on Brouwer's account it is at all possible. To form an opinion about this, we need to understand how, according to Brouwer, expressions or words can receive the peculiar function of evoking mental objects. In particular, should it be assumed that they refer to mental objects? And, if so, are such references set by private ostensive definitions?

Let us first concentrate on the question of reference. Indeed, we may find in Brouwer's papers suggestions that words refer to mathematical constructions, considered as mental. However, these suggestions typically occur in passages presenting his perspective on the issue of the validity of laws of logic. A clear example of this sort is this quote from his (1933) paper:

Will hypothetical human beings with an unlimited memory, who use words only as invariant signs for definite elements and for definite relations between elements of pure mathematical systems which they have constructed, have room in their verbal reasonings for the logical principles as rules for tacking together mathematical affirmations? (*ibid.*, p. 443)

More or less the same question, though with a different wording of the initial supposition, is repeated in (1952, p. 510). He asks the reader to suppose that a "mathematical construction has been carefully described by means of words." However, the two terms: 'invariant signs' and 'careful descriptions' are not univocal enough to construe the initial supposition as assuming that words *refer* to mathematical constructions. Moreover, it is hard to say how much idealization is, according to Brouwer, involved in this supposition. Perhaps it is only an unattainable ideal that words should refer to mental constructions, on a par with the hypothesis of an ideal unlimited memory. Thus, the exegesis of the above fragments is of little help in deciding whether he assumes that words have a referring role.

A clue to this issue can be found in his vision of language as a tool of will-transmission, or more specifically, in the rejection of the view that language represents or describes reality or some fragments thereof. A clear statement of the position that language is not capable of adequately representing or mapping any part of reality can be found not in his own writing, however, but in 'Signifische Sprachforschung' which he co-authored (1919, p. 223). It is claimed there that language is not a medium of representing, but a means by which speakers and hearers influence each other. Instead of the concept of language being a representation of reality, we thus have a picture of command-like utterances by the help of which members of a community influence each other's actions. Now, it is a significant part of this rejected concept of representation that expressions, at least of some categories, refer. Indeed, without the concept of reference, it is hard to see any sense in the idea that language represents. Brouwer clearly rejects the representative role of language, and with it presumably the idea that words refer.

Given that mathematical communication does not imply in Brouwer's doctrine that appropriate words refer to mental constructions, we face a problem of how some utterances are made to evoke the desired mathematical construction in the mind of a hearer. We are still near to the trap of an idea of a privately made decisions as to what constructions a given word should evoke. The following fragment of his dissertation suggests, however, that this was not what he had in mind:

With which mathematical notions a spoken or written symbol will be made to correspond, this choice will take into account as economically as possible the most common mathematical systems and methods of reasoning; therefore it will in general differ according to the milieu. In particular, the answer to the question which domains of mathematics will be accompanied by a language not only among professional mathematicians, but also in daily life, will depend for every nation anew upon the question, which domain of mathematics have found most applications to the guidance of actions or as a means of understanding about action. (1907, p. 73)

Surprisingly, what symbols correspond to may differ, according to the quotation, from milieu to milieu. If this association between mathematical notions and words were effected by an arbitrary decision by each mathematician, we should rather expect that what symbols correspond to might change from mathematician to mathematician. Equally surprisingly, it turns out that the decision as to which part of intuitive mathematics should be expressible in language goes back, in Brouwer's opinion, to the issue of which part of mathematics finds application in the daily life of the community. It thus appears that it is a community, after all, that dictates what language is used and in what way to communicate about mathematical matters.

What remains is to find a picture of how, in Brouwer's doctrine, the subject acquires language within a community. This issue inevitably brings us back to Brouwer's concept of will-transmission. We are offered two pictures, of which the first is, in essence, learning by conditioning. It presents a primitive community, whose members communicate by gestures and cries as they engage in common tasks. A person is taught to obey commanding cries, gestures, and the like. The training he or she goes through aims at evoking "(...) preparedness to labor (...) by giving fright or striking terror, by temptation, by subjugation as one does of animals (...)." (1933, in VS p. 421) The goal is to create in a person a "mathematical view which makes labor acceptable as a means to pleasure or avoidance of pain". (*ibid.*)

One may translate this into saying that as a result of this training an individual conceives of a connection between labor on the one hand and the avoidance of pain or the procurement pleasure on the other. The person becomes aware of what the relation is between his responses to commands and punishments, rewards and reproaches.

The second account attempts to introduce a much more complicated language, as spoken in our linguistic communities. It is important at this stage that each subject assumes as a *hypothesis* that there are other subjects, who have a mechanism of the *mathematical viewing*, acting and reflecting similar to his own. Still another hypothesis is the existence of an

(...) *objective world of space and time* as a common bearer of all the temporal sequences of phenomena of all individuals. (1933, in VS p. 421)

We are told that once labor and the tasks performed in a community became complicated, such simple means of influencing others as gestures, cries and the like turned out to be unsatisfactory. Language comes into being once the subject projects a mathematical system on a totality of commands, gestures, objects, situations, and actions that take place in a common endeavor. The upshot of this is a mathematical system of relations between what we call fragments of external reality and human actions. In the next step the subject gives names to basic elements of the system; grammatical rules also emerge somehow from the mathematical system. This rather obscure story makes one thing clear: the paradigmatic function of language is conveying commands. As Brouwer puts it:

(...) words of present languages (...) in the last analysis are no more than signs of command in society's regulation of labor (...). ('Intuitieve Significa', in VS p. 417)

At first sight, however, it is hard to see how the concept of communication as will-transmission can account for the communication of mathematical constructions. This concept has an application to relatively simple linguistic communication. For instance, if one hears, 'Please bring me a cup of coffee', one understands what the speaker's intention, or *will*, is. It may even be said that for this utterance to be successful, the will of the speaker must be transmitted, since otherwise he may receive milk instead of coffee. In a similar vein, one may try to characterize an exchange of commands and their acknowledgments as they are uttered, for instance, between a bricklayer and his assistant. But what about the descriptive use of language, in which people tell what they have seen, explain what a theory states, or explicate the meaning of a mathematical theorem? It is an essential part of Brouwer's doctrine that the function of will-transmission does not attach to the single utterance on its own, but rather the discourse as a whole is an elaborate net allowing for this phenomenon. This issue is partly addressed in 'Signific Dialogues', where Brouwer takes a stance on Mannoury's opinion that mathematics is practiced for its own sake, independently of social circumstances. In Brouwer's view, a mathematical formula gains its significance from what he calls its asserting character, which is constituted by the formula's being manipulated or operated by members of the community. To illustrate this idea we are offered the following picture of bank bookkeeping:

For the clerks who enter the items in the books, these items have no sense or importance whatever; they do their job to gain a living, but the fact that the board of the bank pays them for book keeping is accounted for by the interests that are served by the transactions. The banker sends an important telegram: this is a living, asserting language; the codewords in which it is written form what I call a sequence of entities which represent another sequence of perceptions and unattained aims, in other words they are as many links between the ultimate aims and the nearest means. If now the content of the telegram is reproduced in the bookkeeping, it does not lose that character but it becomes only more difficult to perceive the connection between means and aims: the sequence of entities is transformed. For the bookkeeper this connection is then dissolved into a general feeling that he conscientiously performed his duty, a feeling that is positively akin to the sentiment that constitutes for the mathematician the notion of 'truth'. (1937, p. 451)

The quotation suggests that basic mathematical expressions and formulae belong to our everyday language. Both basic and more elaborated mathematical expressions evolve and are acquired by a person in social circumstances. Language as a whole is pictured as a network by which various 'will-impulses' are transmitted. Although mathematical constructions are assumed to be mental, there is no essential difference between a way mathematical vocabulary and other expressions are learned.

We have attempted here, rather as an exercise, an interpretation of Brouwer's views of language. Contrary to the received opinion, this analysis defends his account against charges that it is a doctrine of private language. If this analysis is correct, communication in mathematics relies for Brouwer on invoking appropriate mental constructions in the minds of other people. We have argued that the invoking of mathematical constructions need not be thought of as presupposing that mathematical expressions refer. The problem of a word's capacity to evoke a construction in other minds can be handled by pointing to the idea of language as a network of will-transmission and arguing that mathematical expressions are learned in essentially the same way as other sorts of expressions —in social contexts.

5. CONCLUSIONS ABOUT BROUWER'S PHILOSOPHY

To begin with, we have seen that the objection that Brouwer's conception of mathematics does not allow for mathematical results to be communicable is not justified. On the one hand, the assumed cognitive machinery of the knowing subject—causal attention, mathematical intuition of two-ity, mathematical abstraction, as well as the distinction between mathematical relations and their subjective representation—permits the satisfaction of the mentalist condition of intersubjectivity. On the other hand, the doctrine of the mental character of mathematical objects does not, by any means, prohibit satisfaction of the Wittgensteinian condition. It is rather a misguided picture of language that makes one see in Brouwer's doctrine the violation of the above condition. Moreover, one may convincingly defend Brouwer's views on language against the popular objection that they are based on a picture of private language and private ostensive definition.

Our second question concerns Brouwer's semantic views, especially whether they can be seen as offering grounds for rejecting the concept of bivalent truth and replacing it by a rather narrow concept of evidence. That is, we ask whether his demand of 'no non-experienced truths' is a consequence of his account of meaning. However, his reflections on meaning are too scarce to attribute to him this sort of strategy, as they amount to two rather negative claims. He thinks that meaning should be analyzed only in terms of words affecting speakers and hearers and rejects objective senses or concepts. One may also find him saying that some mathematical statements, as construed in classical mathematics, have no meaning. However, this view is rather motivated by his mathematical constructivism and does not follow from his semantic ideas. Such a claim means that if these statements are understood in the intuitionistic manner, that is as requirements to effect appropriate constructions, these turn out to be far beyond anything that can built by following methods acceptable to the intuitionist.

Let us finally try to assess Brouwer's justification for the intuitionistic revision of mathematics. To start with positives, there are, as we have attempted to show, direct connections between Brouwer's epistemology and the crucial ideas of his intuitionism. Thus, the concept of intuition as the sole mechanism for introducing mathematical constructions determines what mathematical objects are allowable. Also, the rejection of some laws of classical logic can be traced back to his views on logic, which assume laws of logic to be idealized regularities obtaining in linguistic reports on the building of mathematical constructions. In this way, two basic tenets of intuitionism, the requirement of constructibility and the new interpretation of the logical constants, are traceable to his philosophical views.

What is more controversial, as it runs contrary to a substantial part of mathematical tradition, is Brouwer's sharp separation between, on the one hand, intuitive mathematics and its creator as well as the ultimate authority on what holds or is certain in intuitive mathematics and, on the other hand, the linguistic representation of mathematics with laws of logic understood as idealized regularities obtaining between reports on mathematical activities of the subject. In particular, mathematical constructions are described as 'languageless', which implies that proofs understood as linguistic objects are not acknowledged as legitimate means of arriving at mathematical truths, though they may, with some accuracy, represent the subject's intuitive procedures for achieving mathematical constructions. In a similar vein, rules and laws of logic are not assumed to be reliable means of proof; they themselves stand in need of justification which is to be sought by examining the relations between assertions of intuitive mathematics. Finally, there is an objection to merely linguistic methods of introducing mathematical objects, such as, for instance, defining a set by means of specifying a predicate that its elements satisfy. As to intuitive mathematics, it is required that its objects be constructed only by means of the intuition of two-ity, which practically implies that they should be arrived at by operations reducible to succession; the intuitive mathematics does not collapse into arithmetic due to the introduction of specifically intuitionistic objects, like choice sequences and spreads, or permission to consider as allowable objects not only outputs of generation of sequences but the rules of their generation as well.

As we said above, Brouwer's program relied on tenets that run contrary to a significant part of mathematical tradition, and were also in conflict with the logistic or formalistic visions of mathematics that prevailed in his day. For this reason, it would be advantageous to restate some of Brouwer's arguments, e.g., against the validity of some laws of classical logic, by appealing to a distinction, motivated by subtlety, between the evidence needed for the assertion of a sentence and the evidence for the assertion of its double negation, and further demanding that valid rules should preserve the evidence instead of preserving bivalent truth. On this construal the issue turns on the existence of convincing objections to such a concept of truth. Our contention is that a persuasive argument of this sort can be erected on two premises, that the only legitimate infinity is potential infinity and that mathematical objects are mental constructions. The argument gains support from the investigation of objects whose becoming is free, that is, choice sequences. We will came back to this argument while examining its more elaborate version. If this reasoning is plausible, the fate of intuitionism depends on the legitimacy of the concept of actual infinity.

To take stock of Brouwer's case, we have an argument for revising classical mathematics that is based on positions, mostly concerning the languageless character of mathematics, that are rather alien to most mathematicians, and therefore of little persuasive power. The argument could be improved on if we had sufficiently strong reasons to reject bivalent truth; given that there is a path from Brouwer's doctrine on infinity to the rebuttal of bivalence, these reasons will turn on its potential character. But, since actual infinity is so well entrenched and so fruitful a notion in classical mathematics, the fate of the intuitionistic project hinges upon how much of classical mathematics it manages to recast in accord with its standards, and how elegantly. As to the philosophical side of Brouwer's doctrine, it should not be accused of various philosophical sins like psychologism, solipsism, advocacy of a conception of private language, appeal to private naming, or violations of the requirement of the intersubjectivity of mathematical results. None of these sins is inherent to Brouwer's conception of mathematics.

3
HEYTING'S ARGUMENTS

There is rather conflicting evidence on how Arend Heyting, probably Brouwer's most faithful disciple, assessed his master's philosophical arguments for the intuitionistic revision of mathematics. On the one hand, he faithfully quotes Brouwer's remarks explaining the concept of the intuition of two-ity, or Brouwer's views on the relation between mathematics and logic. On the other hand, however, the way Brouwer's views are quoted or explained gives the impression that Heyting either holds them irrelevant for the intuitionistic program, or may even be inclined to maintain a certain distance from his teacher's philosophical positions. As we shall see, Brouwer's Kantian approach, which consists in speculation about the epistemic abilities of the knowing subject and consequently which mathematical objects are allowable, is discarded. In a similar vein, Brouwer's concepts of consciousness, mind, causal attention, and the like play no role in Heyting's argument for intuitionism. The notion of mathematical intuition receives a new reading. The concept of number is introduced in psychological terms and related to a faculty that serves to discern *entities* in somebody's mental content. Furthermore, Brouwer's overall negative appraisal of logic is replaced by a much more liberal stance that allows for investigations of the logic of intuitionistic mathematics. There is a considerable shift of attitude towards philosophy as well. Heyting claims that "no philosophy is needed to understand intuitionistic mathematics" and that intuitionistic mathematics is simpler than any philosophy. (1974, p. 79) He therefore does not attempt to justify the intuitionistic revision philosophically. The superiority of intuitionistic mathematics is argued on the grounds of its simplicity, higher subtlety or even its independence from controversial philosophical assumptions. As a result of these changes, much of the militant spirit that was associated with Brouwer's attack on classical mathematics is gone. They also suggest that coming to Heyting's arguments for intuitionism is like entering a new landscape. Let us study it in more detail.

1. AGAINST INTUITIONISTIC PHILOSOPHY, FOR INTUITIONISTIC PSYCHOLOGY?

Heyting's late overview of intuitionism (1974) begins with an explicitly stated separation between intuitionistic mathematics and issues of a philosophical nature. While addressing the questions of

(...) why mathematical theorems are so certain and whence mathematics takes its evidence and its indubitable truth (...) (1974, p. 79)

he answers that

(...) basic notions of mathematics are so extremely simple, even trivial, that doubts about their properties do not rise at all. (*ibid.*)

For this reason, intuitionism is not to be compared with a philosophical position like realism or idealism. This is so because

(...) every philosophy is conceptually much more complicated than mathematics. (*ibid.*)

He summarize this rather positivistic stance with a *dictum*:

The only philosophical thesis of mathematical intuitionism is that no philosophy is needed to understand mathematics. (*ibid.*)

As far it goes, this slogan is certainly true; however, Heyting seems to go further. The impression arises that, in his view, no philosophy is needed to *justify* the intuitionistic revision of mathematics. Clearly, this stance is in opposition to his master's enterprise, as metaphysical and epistemological premises played an important role in Brouwer's arguments. More precisely, Heyting believes that these assumptions are not inherent in the intuitionistic doctrine, and suggests that it would be better for the sake of intuitionism to eliminate them. Commenting on Brouwer, he complains that

By clinging to a Kantian starting point he has not escaped the danger that these premises may be considered as essential for intuitionism. (1958b, p. 102)

One may discern two strands in Heyting's attempt to purge intuitionism from philosophy. One is a rather evident perception that to practice intuitionistic mathematics one need not know of, or subscribe to, any philosophical doctrine. The other maintains the philosophical neutrality of the intuitionistic *program* of reconstructing mathematics and contrasts this with classical mathematics that allegedly must introduce metaphysical assumptions in order to justify the forms of reasoning it applies. Thus, according to Heyting, the superiority of intuitionistic over classical mathematics consists in the former's being free of metaphysics while the latter is based on the Platonist presupposition that mathematical entities exist timelessly and independently of the human mind. Since that assumption, being philosophical, is far from obvious, intuitionistic mathematics is more evident and secure than its classical rival. Thus, in order to be attracted by the intuitionistic criticism of classical mathematics, one needs to become

(...) aware of the fact that the idea of natural numbers (and other mathematical entities) as existing independently of their mental construction is not sufficiently clear to serve as a basis for mathematics. (*ibid.*)

The hypothesis of the independent existence of mathematical objects is, according to Heyting, an essential ingredient of classical mathematics, as it is necessary for the *justification* of means of proof of classical mathematics, like for instance the rule of double negation elimination or the principle of the excluded middle.

It is surprising to find that Heyting's position is akin to positivism. It says that we have the right to study mathematical structures only as they present themselves to our minds; our methods of deriving mathematical truths should be tailored to the way these structures are given. Classical mathematics is accused of attempting to study mathematical structures 'in themselves', far beyond how they are presented to us. Remembering Brouwer's metaphysical extravagance, it is somehow sobering to find Brouwer's student arguing for intuitionism on the ground of its (alleged) philosophical neutrality. Together with Heyting's rejection of philosophical speculations goes a demand that the basis of mathematical certainty should be found in something immediately given and understandable, which does not involve philosophical subtleties. (1974, p. 79) Such a basis is found first in the process of counting. But the counting of material objects, as it presupposes both the external world and abstract numbers, is also, according to Heyting, not philosophically neutral—it cannot serve as a foundation of mathematics. He turns then to more immediately given phenomena, found in the faculty of isolating an object. Processes of isolating allow the subject to discern constituent entities in the field of perception, the latter being assumed to be always given as an undifferentiated whole. To take Heyting's example of visual perception: by focusing one's vision in a certain direction, one receives a mental content which is given as a whole. (*ibid*. p. 80) Then, by concentrating or fixing our attention on a particular impression (a group of impressions) the subject isolates or forms an object. This faculty is claimed to be a fundamental function of human mind, without which no thinking is possible. Instead of explaining the faculty of isolating an entity in terms of *focusing* or *fixing* our attention, Heyting introduces a more Brouwerian term, namely *mentally creating an object*. (*ibid*.) For processes of counting, moreover, what the created objects are is inessential. Rather, it is merely the act of isolating and the resulting *one* isolated object that serves as the basis of mathematics. Thus, he talks of the ability to abstract an isolated object from perceptual content. Both faculties are essential for human thinking (not only mathematical). As Heyting states, "When we think, we think in entities." (*ibid*.) Later however, a qualification is added to the effect that not all mental life consist in thinking in terms of entities since this quality fades away when we live intensely, and "under the influence of strong emotions the world seems a whole." (*ibid*.) To a philosopher who may at this point ask how it is possible that human beings 'think in entities', Heyting replies that this problem need not be addressed in mathematical considerations as it belongs to philosophy proper. It is simply a fact that we think in entities. Then, after having addressed the issue of the origin of the concept of *one*, he proceeds to such problems as how an entity is recognized as identical and distinct from other entities. Again, he refuses to follow a speculative approach to these questions. Instead, he states that it is simply a fact that we are able to fix our attention on a perception, and then, while retaining it in the memory, isolate another entity. We are also able to abstract from features of created objects and consider them as 'pure entities'. By repeating these processes further we construct arbitrary natural numbers. (*ibid.*, p. 81)

Despite Heyting's professed intention of purging foundational studies from philosophy, a careful reader may note quite a few Brouwerian themes in his investigations of the basis for mathematics, which we sketched above. Where Brouwer talks about mind and consciousness, Heyting makes a distinction between thinking in entities and experiencing the whole of a mental content, without discerning constituent entities. Instead of Brouwer's concepts of the intuition of two-ity and the causal attention we have in Heyting an account of the faculty of isolating an entity together with an ability to fix attention on still another perception. While Brouwer introduces another faculty of the mind—mathematical abstraction—Heyting simply talks about our ability to abstract from the contents of perceptions. However, despite these similarities of themes, the accounts differ diametrically. As Heyting would have it, all the abilities are interpreted as psychic, that is, they are faculties of flesh-and-blood persons, and not of an abstract knowing subject. It is likely that he read a psychological doctrine into Brouwer's highly speculative philosophy. For instance, speaking about Brouwer's distinction between mind and consciousness, he states:

Evidently, this distinction is based on a psychological theory concerning the origination of knowledge. (Heyting 1958b, p. 101)

Thus, processes like focusing attention, thinking in entities, retaining something in memory, and abstracting from the content of a perception are psychological facts for Heyting. This is a considerable shift of attitude since, as we have argued before, the most tenable interpretation of Brouwer's system is to assume that the knowing subject is *transcendental*, that is, it should be understood more or less along Kantian lines, and consequently its abilities should not be seen as psychic faculties of a flesh-and-blood mathematician. An undeniable advantage of Heyting's approach is that it puts the fate of intuitionism on a more 'down-to-earth' ground, so the potential convert to intuitionism is not distracted by the burden of speculative philosophy. Its more problematic aspect is its appeal to psychology or psychic facts, which are believed capable of serving as a 'basis for mathematics'. Indeed, one may find close similarities between his account and attempts to 'found' mathematics on psychological investigations, which were proposed at the end of 19th century. One may wonder how such investigations can disclose a basis for mathematics, although it may be hoped that once Heyting's 'basis' is appropriately understood, psychologism will disappear from his doctrine. More importantly, it is notoriously unclear how psychological reflections may show that intuitionistic mathematics is correct, that for instance the logical connectives should be interpreted in the intuitionistic manner. Still more obscure is how an argument for intuitionistic revision of mathematics may follow from such considerations.

At this point it may be asked what prevents Heyting from taking a more standard approach to the foundations of mathematics, that is, from selecting some logical laws and rules and mathematical axioms as the starting point of a mathematical theory. To this proposal he replies, somewhat surprisingly, that logic is not neutral in regard to philosophical issues. As the crucial logical concept is that of a *proposition being true*, we encounter at once the problem of explaining what proposition and truth are. Thus, he asks somewhat rhetorically

> But what is a proposition? Does it coincide with the sentence by which it is expressed or is it something behind the sentence, some meaning? If so, what is the relation between the proposition and the sentence? And what does it mean that the proposition is true? Does this notion presuppose the existence of an external world in which it is true? (1974, p. 79)

And he goes on to remark that these questions have been answered differently, in a number of ways, none of which is convincing enough. Consequently, he draws the conclusion that logic is too problematic to serve as a basis for mathematics. In other words, since he sees logic as philosophically loaded, he discards the logicist's approach. Besides, there is also a second argument of Heyting's against a logical foundation of mathematics that goes back to Brouwer's views on the relation between mathematics and logic. Brouwer's stance is considerably moderated, as no appeal is made to the doctrine that identifies laws of logic with idealized regularities obtaining among utterances. Instead, they are considered to be mathematical theorems of the highest generality, and for this reason, less evident than other mathematical statements. Because of this, they cannot serve as a foundation of mathematical theory. (*ibid.*, p. 79) At this point one may object by saying that Heyting confuses questions of how evident mathematical theorems are and how rich in consequences they are. Neither Frege nor Russell believed that statements they selected as logical axioms and from which the whole of mathematics was to be derived were more evident than simple arithmetical equations, or theorems about natural numbers. Apart from a hierarchy of theorems built in according to their evidence, they supposed, however, that some statements are more basic than others, because they are richer in consequences.

In Heyting's case for intuitionism, two sorts of arguments (or perhaps, elements of arguments) can be discerned. One is a psychological reflection on the origination of mathematical concepts; I shall dub it the 'psychological argument'. The other amounts to a thesis that purging mathematics of philosophical premises will lead to the replacement of classical by intuitionistic mathematics. The premise questioned is the Platonist's assumption of the independent existence of mathematical objects. This argument attempts to establish that from a postulate that mathematical entities should be studied as immediately given to the mind, without supposing that they exist in this or that way, a revision of logic and mathematics follows. I will call this reasoning the 'neutrality argument'. There is also a semantical argument, loosely connected with the previous one, whose starting point is a reflection on meaning of mathematical statements, and which attempts to show that once intuitionistic meaning is ascribed to mathematical statements, it will lead to the intuitionistic interpretation of the logical connectives and quantifiers. We shall discuss these arguments below.

2. INTUITION AS SELF-EVIDENCE

As we argued in Chapter 2 section 2, the concept of intuition plays a crucial role in Brouwer's reconstruction of mathematics. It is called the 'intuition of two-ity' since, as the empty form of all change, it lies behind any passage from one sensation to another. The possibility of thinking such acts as being indefinitely repeated gives rise to a construction of arbitrary natural numbers. Moreover, two basic methods of introducing mathematical objects, that is, either as choice sequences or as spreads, rest upon the progression that this intuition guarantees. Heyting finds a 'basis for mathematical truth and certainty' in the psychological faculties of isolating an object, the possible repetition of this process with still another object, and abstraction from a mental content. But what is meant by saying that something is a basis for mathematical truth? On the one hand, his psychological account can be interpreted as a rather innocuous and non-controversial doctrine about the *origination* of the concept of natural number. On the other hand, however, one may read it as a more far reaching explanation of the nature of mathematical *truth*, which attempts to answer what it means that a mathematical statement holds. On the former reading, the account does not offer any explanation of how one introduces and then grasps more elaborate mathematical concepts, apart from a rather empty stipulation that these should somehow be built out of 'evident' ones, that is, out of natural numbers. It is, however, the latter reading that is troublesome since it is highly doubtful how alleged psychological abilities to think in terms of entities, isolate objects, and abstract from mental content can have a bearing on the question of whether a mathematical statement holds. Thus, one should rather read Heyting's remarks about the basis for mathematics as concerning the justification of mathematical theorems. They suggest that the mathematician recognizes whether or not a construction obtains by consulting the faculties mentioned. In other words, this psychological account postulates an epistemic ability that allows a mathematician to recognize that a mathematical proposition holds or that a mathematical construction is correct. Such an ability may be called 'intuition' although Heyting only rarely uses this term. Contrary to Brouwer's idea that intuition is some means of building mathematical constructions, Heyting's concept is akin to a traditional view of intuition that regards it as a direct and non-inferential means of seeing that a proposition holds. He explains thus intuition as a faculty that "puts mathematical concepts and proofs immediate before eyes." (1934, p. 388) While characterizing Brouwer's standpoint, he insists that intuition is not to be identified with an ability that provides, rather mystically, an insight into the world. It is rather a faculty that allows one to "consider separately certain concepts and proofs that regularly occur in everyday thinking." (*ibid.*, p. 388) One can hardly recognize in these remarks anything similar to Brouwer's teaching on intuition. What is also new in Heyting's conception is a hint that intuition is not infallible. While considering the question of whether absolute rigor and certainty are realized in intuitionistic mathematics, he replies that "absolute certainty for human thought is impossible and even makes no sense." (Heyting 1958b, p. 103) This may suggest that intuition does not guarantee that a mathematical construction recognized as intuitively correct need be correct in fact. Where shall we look then for a justification of intuitionistic mathematics? Heyting's answers is: "it lies in the immediate conviction of the self-evidence of such propositions as this, that if a set contains 5 different elements, it also contains 4 different elements." (*ibid.*, p. 103) Thus, Brouwer's intuition of two-ity is replaced by a conviction, or recognition, of self-evidence. Since self-evident propositions are the propositions about (small) natural numbers, he calls his concept of intuition: the intuition of number or the intuition of natural numbers.

Clearly, the attempt to found mathematics on a postulate of the self-evidence of its propositions or on self-evident statements about natural numbers gives rise to a number of objections. To list two of them, one may convincingly argue that self-evidence relates, first of all, to a *mathematician's* capacity to see a theorem or its proof as obvious, trivial, or self-evident. Consequently, self-evidence is not an absolute property. It may vary from mathematician to mathematician. Secondly, one may notice that many intuitionistic proofs are rather far from being evident (e.g., proofs in intuitionistic analysis). Moreover, these proofs are frequently much less evident or simple than their classical counterparts. More important in the history of intuitionism than any of these objections, however, was Griss's denial of the applicability of negation in intuitionistic mathematics. Griss started from the perception that, in order to carry out a construction appropriate for the assertion of a negated statement, one needs to present in intuition (or grasp as evident) constructions that are mathematically impossible. His criticism of using negation in constructive mathematics assumes Heyting's teaching on intuition and evidence and then derives a result incompatible with Brouwer's and Heyting's mathematics.

It appears, however, that Griss's attack only indirectly concerns Brouwer's philosophy of mathematics, in which it is not required that constructions should be evident. Moreover, at least in Brouwer's earlier papers, negated statement $\neg p$ is explained as asserted on the ground of the subject's observation that "(...) the construction [required for asserting p] no longer goes (...)" (1907, p. 72) and accordingly, there is no mention of an operation on the impossible construction, that is, the operation of deriving a contradiction from the assumption that the impossible construction is given.[38]

To begin with, let us note first that for Heyting a mathematical statement is an expression of the successful carrying out of an appropriate construction. (1958a, p. 333) To take his own example, the statement '6 is even' expresses a construction that consists in finding such a natural number that, if multiplied by 2, yields 6. Clearly, the required number is 3. But what about the statement '5 is even'? As the construction of such a number is impossible, it is doubtful whether we can know what this construction is, or whether we could recognize it. Nevertheless, we need to have a concept of such a construction since, by intuitionistic standards, accepting a negated statement like 'It is not the case that 5 is even' requires a method showing that any construction of an even number 5 is impossible or leads to a contradiction. Thus, in order to assert the negated statement $\neg p$ one needs first to form a conception of what construction entitles the assertion of p, that is, to present in intuition an appropriate construction for the false statement p, and then to show how any such construction leads to a contradiction. A crucial question concerning this doctrine is how one can envisage or present in intuition a construction that is clearly impossible. Is there any sense in the claim that an impossible construction is evident or intuitive? Commenting on Griss's argument, Heyting remarks that the conclusion of his reasoning is that a false statement cannot be meaningfully uttered and consequently negation cannot be used at all. (*ibid.*, p. 333)

Heyting has a twofold reply to Griss's objection. First, he accuses his line of argumentation of operating with too narrow a concept of construction that assumes as legitimate only 'completed' or 'ready-given' constructions (*vollendete, fertig vorhandene*). (*ibid.*) On the contrary, he argues, constructions considered in intuitionistic mathematics only rarely have this character. It is clear that a construction showing that a simple numerical statement like $7 + 4 = 11$ holds is 'completed' or 'ready-given'. However, when we come to a further stage and consider a statement like: 'For any n, $n \times 2 = 2 \times n$', what justifies it is not a construction, but rather a *method* of construction, which when applied to a given number a (or more precisely, to a construction of a given number a) yields an equality: $a \times 2 = 2 \times a$. The example shows that even in elementary intuitionistic mathematics there is a need to operate with constructions of varying complexity. Instead of introducing a distinction between *method* of construction and *construction* itself, one simply concedes that some constructions operate on other constructions, say subsidiary ones, while others do not. And, as Heyting sees Griss's argument, it is the legitimacy of a subsidiary construction that is objectionable. In accord with intuitionistic standards, the statement 'It is not the case that 5 is even' should be translated as:

As soon as I have a construction B of a natural number n such that $5 = 2 \times n$, I can derive a contradiction. (*ibid.*, p. 334)

The construction B is, however, impossible or in Heyting's terminology, *non-realizable*. Obviously, this new terminology does not help to answer Griss's objection; he will still ask how non-realizable constructions can be evident or presented in intuition. And if they are not evident, one should exclude them from intuitionism, the consequence being that negated statements of intuitionism should be translated into 'positive' ones. To block this reasoning, Heyting declares that in intuitionistic mathematics one uses various constructions: completed, incomplete but in principle realizable as well as non-realizable. He then goes on to insist that evidence is gradable, that there are 'stages of evidence'. The most evident are constructions justifying statements about small natural numbers, like 2+2=4; he calls them 'unconditional and immediately surveyable'. (1958a, p. 335) However, coming to larger natural numbers the highest evidence is gone—the construction of 1000+2 = 1002 is less evident than that of the former. Still less evident are constructions justifying universal statements concerning natural numbers, like

For any n, $n \times 2 = 2 \times n$.

For, in this case one is dealing with a construction operating on other constructions, which is not considered to be completed. At the next stage we encounter methods that make use of impossible or unrealizable constructions. Somewhat surprisingly, aside from the hardly evident impossible constructions, we have in Heyting two more stages of lesser evidence. First, less evident are constructions justifying statements about various *species* of natural numbers, like for instance this: if $A = B$ and $B = C$, then $A = C$, where A, B, and C are species of natural numbers. Still less evident are constructions occurring in the mathematics of *spreads*.

At first sight the theory of 'stages of evidence' appears to be psychologically credible and hardly controversial. Obviously, equalities obtaining between natural numbers present themselves as far more evident than propositions concerning spreads, for instance. It is not clear, however, what this account is meant to accomplish. Does it claim that for any intuitionist, his estimation of the degree to which constructions are evident should be in accord with the above hierarchy of evidence? Heyting concedes that this is a subjective matter: people may differ in their evaluation of what is less and what is more evident. (*ibid.*, p. 337) For somebody hoping to find a basis for mathematics in the conviction of self-evidence, this is not an encouraging state of affairs. Perhaps one should only assume that simple arithmetical constructions are self-evident and argue that the rest of intuitionistic mathematics should be built on or derived from them. This is the opinion that Heyting expresses in a few papers. For instance, he argues that natural number is a basic notion of mathematics since it is "easily understood by any person who has a minimum of education"; numbers are "universally applicable in the process of counting" and the notion "underlies the construction of analysis." (Heyting 1956, p. 15) An implicit suggestion is that all other notions and constructions should somehow be derived from the notion of natural number. However, the idea of building or deriving intuitionist mathematics from arithmetic has a notoriously unclear meaning in this context. One finds out what this idea means if one studies, for instance, how the intuitionistic theory of real numbers is developed from arithmetic; for ideological reasons, both Brouwer and Heyting refuse to list what methods are permitted in this derivation, as they fear that this would limit the freedom of mathematical thinking.[39] We are thus left with problems like 'why Dedekind's theory of real numbers is claimed not to be built or derived from basic notions of arithmetic, but Brouwer's theory is believed to be so'. On this account of Heyting, the intuitionistic repudiation of classical mathematics appears to be arbitrary, or at least, not justified by arguments appealing to intuition and evidence. The argument that appeals to intuition understood as a conviction of self-evidence can hardly succeed in convincing classical mathematicians of the incorrectness of their discipline. Moreover, the subjective character of this notion is hard to reconcile with the idea that what is true or certain in intuitionistic mathematics does not vary from intuitionist to intuitionist. Likewise it is hard to believe that we do not differ in our estimations of what is evident. Thus, it is much safer to say, like Brouwer, that the mechanism of one sensation giving way to another is the same in any thinking subject. For these reasons, Heyting's concept of intuition is a step backwards in relation to what was introduced by Brouwer.

That this appeal to intuition is far from being satisfactory was likely felt by Heyting himself, when he conceded that "it has proved not to be so intuitively clear what is intuitively clear in mathematics." (1962, p. 195) That is, even intuitionists may disagree about which parts of mathematics are less, and which are more, evident. But he maintains that contrary to these possible differences of opinion, they agree that the Platonist assumption of mathematical objects having existence independent of thinking subjects cannot play any role in mathematical reasoning. This is not only because of the requirement that mathematics be separated from philosophy, but also because such an assumption is by no means evident or given.[40]

3. THE NEUTRALITY ARGUMENT

Heyting presents intuitionism as a philosophically neutral way of practicing mathematics: the way this mathematics is practiced and the methods it uses do not presuppose adherence to any philosophical thesis. Similarly, he claims that these methods do not stand in need of philosophical justification. Both of these professed virtues of intuitionism are contrasted with classical mathematics, as he holds that the means of proofs accepted in the classical school *do* depend on the metaphysical doctrine of Platonism. Thus, if we can manage to show this difference between the two schools, we will have a relatively strong argument for intuitionism on the grounds of its not appealing to a controversial metaphysical thesis. But, in order to establish this claim, Heyting's argument must prove two things. First, it should show that classical means of proof stand in need of justification and that justification must be based on mathematical Platonism; consequently, the repudiation of this metaphysical doctrine would lead to the rejection of the forms of reasoning of classical mathematics. Secondly, the argument needs to show that purging mathematics of this metaphysical assumption leads exactly to the methods accepted in intuitionism. The key word here is 'exactly' since, hypothetically, if this purging led, for instance, to giving up mathematical induction, then the argument would work against intuitionism. Although Heyting's project is quite general, his argumentation and, especially, his examples, are fairly limited in scope. First of all, the methods claimed to be dependent on metaphysics are identified with the laws of logic. That is, the argument attempts to show that the laws of classical logic stand in need of metaphysical justification, and moreover that without such justification they are arbitrary. Secondly, although this characterization of the project may give the impression that it aims quite generally at identifying and then removing from mathematics *any* philosophical assumption, it concerns only the metaphysical thesis mentioned, that there are mathematical objects existing independently of whether or not a mathematician thinks or may think them. With these two qualifications, the argument amounts to showing that once mathematics is cleansed of Platonism, some laws of classical logic will cease to be universally valid. Against Platonism, Heyting holds that mathematical objects can only be studied as given to, or constructed by, the subject. In turn, the claim that *there is* a mathematical object x should be understood as a claim that x is constructed, or rather, that an effective method for its construction is known. Consequently, one may look at Heyting's 'neutrality-argument' from an ontological perspective, as attempting to draw a conclusion about the validity of the laws of logic from the assumption that mathematical objects are mental constructions.

3.1. *Heyting's Counterexamples: What Do They Prove?*

To begin with the neutrality argument, let us consider one of the best known of Heyting's examples that aim at exposing the Platonist underpinning of the laws of classical logic. Let us look into the following definitions of some natural numbers:

k is the greatest prime such that $k-1$ is also a prime, or $k=1$ if such a number does not exist.
l is the greatest prime such that $l-2$ is also a prime, or $l=1$ if such a number does not exist.
(Heyting 1956, p. 2)

Now, the intuitionist says that the fact that both definitions are syntactically much alike tends to blur the significant difference between them. The former allows us to determine what the defined natural number is, while the latter does not permit us, in the present state of knowledge, to calculate l. At this point the classical mathematician may ask why one should bother about this difference. The form of the definition of l may be represented as follows:

$$l = \begin{cases} n & \text{if } \psi, \\ 1 & \text{if not-}\psi, \end{cases}$$

where ψ expresses a yet-undecided proposition and n is a certain natural number, completely determined by ψ if only ψ holds. Given this form, the assertion that there is l does not lead to a contradiction. Moreover, on the assumption of the principle of non-contradiction, as it cannot be that both ψ and not-ψ hold, the intuitionist must agree that, if there is l, then it is unique. Thus, the classical mathematician may say that both requirements for correct definition, i.e., the non-contradictority and uniqueness of the defined object are satisfied and that what is being disputed with the intuitionist is not the validity of the laws of logic, but the question of whether or not these requirements suffice. Intuitionists require that a method of calculating the object defined be known, which in the case of natural numbers means that the method should allow us to state what the number in question is. In the case of real numbers, this means that a defined number can be calculated with an arbitrary exactness: for instance, for any place of its decimal expansion the method should permit us to calculate the numeral that occupies that place. It is plain that this postulate is a consequence of Heyting's general thesis that mathematical objects should be studied as given, or constructed.

However, against the claim that l is *not-constructed* one may issue the following objection. Clearly, by intuitionistic standards the number 1 *is* constructed. Similarly, if you only allow for an idealization (to which intuitionists do not object), any arbitrarily large natural number is also constructed. Accordingly, if there is the largest twin number, then it is identical to a number that has already been constructed. Thus, the objection goes, the number l is constructed as well, since it may be identical with only one of several numbers, each of which is constructed. This argument is, however, incorrect, as the phrase 'is constructed' is not extensional, even in the classical mathematician's construal. To see this, suppose a constructive proof of $\psi(a)$ is known, and a is known to be identical with b, but only due to an indirect proof. These two pieces of information do not allow us, in general, to obtain a constructive proof of $\psi(b)$. Similarly, the fact that any given natural number is constructed does not prove, in the lack of a constructive proof that l is identical with some natural number, that l is constructed as well. Things would stand differently if we restricted ourselves to constructive identity only, that is, assertible only on the grounds of the knowledge of a constructive proof that the identity holds. However, it is just the point of Heyting's example that we do not know of a constructive proof that l is identical with a given natural number.

Suppose that at this point the classical mathematician provisionally accepts the intuitionist's reservations towards the definition of l and asks for a demonstration of why the rejection of this definition exhibits the invalidity of the principle of the excluded middle. Clearly, his diagnosis of the argument is that all laws of classical logic hold and the intuitionist is imposing only an additional demand on the correctness of definition, over and above the uniqueness of the defined object and the definition not being contradictory. The intuitionist is bound to retort that their divergence of opinion is not simply about the strengthening of requirements for the correctness of definitions, but rather concerns the issue of how the statement ψ that figures in the definition should be understood. Moreover, he would add that the classical interpretation is utterly confused as it invokes a picture of the *complete* totality of twin numbers, of which it is meaningful to suppose that it either has a given property or not, no matter what our knowledge on this subject is. The point of contention is not that the sequence of twin numbers is ill-defined: it is the intuitionistically legitimate sequence of natural numbers subject to the condition that its n-th element is the n-th twin number. (We follow here the convention of reserving the name 'twin number' for number k, given k and $k-2$ are both primes.) The sequence may or not come to a stop and we can assert one or the other of these possibilities if we prove it on the basis of our knowledge of how the generation of that sequence proceeds. More generally, a disjunctive property, A or B, can be asserted of a sequence if we are in possession of a proof that a sequence has the property A, or a proof that it has the property B.

The classical mathematician may agree that given how the intuitionist interprets mathematical statements, he cannot in general assert that $\varphi \vee \neg\varphi$ holds; however this fact has nothing to do with the validity of the principle of the excluded middle since the above claim, if properly translated, amounts only to the statement that 'A proof that φ is known or a proof that not-φ is known' and this clearly is not an instance of the principle mentioned. Thus, to arrive at the desired conclusion, the intuitionist must show that his rival misconstrues mathematical statements, as his interpretation of them must rely on Platonism. It is rather plain that the acceptance of complete infinite totalities provides a strong motivation for accepting the principle of the excluded middle, but our problem is the opposite. Must the classical mathematician be committed to complete infinite totalities in order to claim that the principle holds in reasoning pertaining to infinity? He may say that the excluded middle holds in arithmetic since the way natural numbers are introduced, as generated by the succession operation, determines of any property whether or not it holds, no matter whether natural numbers are thought of as a complete totality, or as an infinitely proceeding sequence. This debate deserves our special attention but as it has much to do with the topic of the next chapter, we shall postpone the discussion until later.

A similar moral to the effect that intuitionists urge a new interpretation of mathematical statements rather than supplying counterexamples to the laws of classical logic can be drawn from Heyting's reasoning directed against the validity of Dedekind's definition of real number. In essence, the definition relies on ascribing to any rational number one of two predicates, *left* or *right*, in a way that accords with the natural ordering of rationals. A real number is given by a sequence of nested intervals, each containing the defined number, such that their lower ends have the property *left* while the upper ends the property *right*. From the intuitionistic perspective, however, this idea is incorrect, as it does not allow us to decide, for any rational number, whether it lies to the left of, to the right of or is identical with the defined real number. (Heyting 1931, p. 107) Accordingly, the intuitionist considers this method to be metaphysical, as it relies on assuming objective order relations that obtain, no matter whether or not one has a method of identifying which ones hold. The classical mathematician asserts that for a given real number t and any rational number x, either $t = x$, or $t < x$, or $t > x$. For the intuitionists, the assertion of each of these disjuncts should be justified by the appropriate construction. A real number is given by a sequence of nested intervals that comprise that number, and whose endpoints are rational numbers. To claim that $t < x$ one needs to find such an n that the upper end of the n-th interval is smaller than x, to assert that $x < t$ one must find a k such that x is smaller than the lower end of the k-th interval. Finally, to assert $t = x$ one needs to show that for any n, x is smaller than or equal to the upper end of the n-th interval and is greater to or equal to its lower end. Thus, the classical mathematician holds that either there is a natural number N such that after N steps of the calculation of t it turns out either that $t < x$ or that $x < t$, or no such N exists. Since for some numbers, commonly believed to be real, none of the three constructions is known, the intuitionist concludes that by Dedekind's definition they are not real numbers. That is, Dedekind's definition, if understood in accord with intuitionistic standards, does not permit the assertion that such numbers are real. The counterexamples to Dedekind's method are provided by those reals whose order relation to some rational number is not known. Heyting's σ-number of which it is not known whether or not it equals 1/3 shows how to construct such counterexamples with the help of a yet-undecided mathematical problem. On the other hand, as Brouwer and Heyting show, to establish order relations between an irrational number and an arbitrary rational number, it suffices to have a proof of the irrationality of that number: such a proof provides an estimation of that number with respect to an arbitrary rational. (Heyting 1956, p. 29) However, while no proof of the irrationality of some numbers, for instance Euler's constant C, is known, there are no grounds for believing that for any rational x, it can be proved whether $x < C$ or $C < x$ or $x = C$. As in this critique it is denied that the disjunction $C = x$ or $C < x$ or $C < x$ holds, Heyting adds the gloss that the principle of the excluded middle cannot be applied there. (1931, p. 108) It is clear, however, that the point of the dispute is rather how to interpret sentences concerning order relations between real numbers. Given that they are construed along intuitionistic lines, the disputed disjunction is no longer an instance of the classical principle of the excluded middle; hence whether it holds or not has no bearing on the question of the validity of the laws of logic. This brings us to the issue of how the intuitionist thinks of order relations.

For the intuitionist a real number is given by a real number generator, that is by an intuitionistically interpreted Cauchy sequence:

A sequence $\{a_n\}$ of rational numbers is called a *Cauchy sequence* if, for any natural number k we can find a natural number $n = n(k)$, such that $|a_{n+p} - a_n| < 1/k$ for every natural number p. (Heyting 1956, p. 16)

To introduce real numbers one defines first the relation of coincidence, $=$, between generators:

For real number generators $a \equiv \{a_n\}$ and $b \equiv \{b_n\}$, $a = b$ if for every k one can find $n = n(k)$ such that $|a_{n+p} - b_{n+p}| < 1/k$ for every p. (*ibid.*)

Finally, a real number is defined as a species of infinite sequences of rational numbers that coincide with a given real number generator. The order relation between reals is then introduced in terms of the ordering of generators:

For real number generators a and b, $a < b$ if one can find natural numbers n and k such that $b_{n+p} - a_{n+p} > 1/k$ for every p. (*ibid.*, p. 25)

Now, given these definitions, it may happen that the intuitionist cannot assert, of two real numbers x and y, that either $x = y$ or $x < y$ or $y < x$. Just consider two reals given by the following sequences:

$\varphi_n = \frac{1}{3}$ for all n;

$$\sigma_n = \begin{cases} \frac{10^n - 1}{3 \times 10^n} & \text{if } n \text{ is such that in the decimal expansion of } \pi \text{ the digit 9 proceeded by the sequence } 0,1,2,3,4,5,6,7,8 \text{ does not occur before the } n\text{-th place} \\ \frac{10^k - 1}{3 \times 10^k} & \text{if } n \text{ is such that at the } k\text{-th place, } k \leq n, \text{ of the decimal expansion of } \pi \text{ the digit 9 proceeded by the sequence } 0,1,2,3,4,5,6,7,8 \text{ occurs.}^{41} \end{cases}$$

It is easy to check that $\{\varphi_n\}$ and $\{\sigma_n\}$ are real number generators. The real numbers they introduce are denoted by φ and σ, respectively. We may ask thus which of the above statements holds: (1) $\varphi = \sigma$, (2) $\varphi < \sigma$, or (3) $\sigma < \varphi$. For (1) to hold, the method of finding, for any k, a natural number $n = n(k)$ should be known, such that $|\varphi_{n+p} - \sigma_{n+p}| < 1/k$ for every natural p. This requirement, however, is satisfied only if we have a proof that the sequence $0,1,2,3,4,5,6,7,8,9$ does not occur in the expansion of π and. Since the proof is not known, it is also not known if for any k the number n can be found such that the inequality holds. Accordingly, (1) cannot be asserted in the present state of knowledge. Secondly, the way φ and σ are defined assures that (2) does not hold, either. Thus, the remaining possibility is (3) which obtains if the method of finding such natural n and k is known, that $|\varphi_{n+p} - \sigma_{n+p}| > 1/k$ for every natural p. But, to know this method, it is necessary to know that somewhere in the expansion of π the sequence $0, 1, 2, 3, 4, 5, 6, 7, 8, 9$ occurs. Accordingly, (3) also cannot be asserted. To summarize, the classical mathematician sees in the disjunction '(1) or (2) or (3)' an instance of the principle of the excluded middle; the intuitionist translates the disjuncts into claims of the form 'The method x is known', which is in accord with his anti-metaphysical stance. As a result the above disjunction cannot be seen as an instance of the principle of the excluded middle.

A similar translation is operative in Heyting's argument designed to impair the validity of the rule of double negation elimination. To this end, let us consider again the real number σ defined by the generator $\{\sigma_n\}$. Now the question is whether σ is rational. One has the right to assert the statement 'σ is rational' only if he knows a demonstration that σ is equal to a fraction k/l. But in order to achieve this demonstration we need to know whether, and if so where, the sequence $0, 1, 2, 3, 4, 5, 6, 7, 8, 9$ occurs in the expansion of π. As this is not known, it cannot be asserted that σ is rational. But could we assert that it is not the case that σ is not rational? Given intuitionistic negation, 'σ is not rational' means that from the assumption that it is rational, a contradiction follows. Let us thus assume that σ is not rational. This has the consequence that σ cannot be equal to $\frac{10^k-1}{3 \times 10^k}$ for any k. Accordingly, $\sigma = \frac{1}{3}$, which contradicts our assumption, since clearly $\frac{1}{3}$ is a rational number. Heyting concludes that from the fact that it is not the case that σ is not rational it does not follow that σ is rational, which at first sight looks as if the rule of double negation elimination were being rejected. However, once we reflect on the way the intuitionist interprets the sentences involved, it will turn out that the example fails to impinge on the rule mentioned. By intuitionistic standards, the double negated sentences read as a claim that one has a method of deriving a contradiction from the assumption that a proof is known that a contradiction follows from supposing that σ is rational. In turn, 'σ is rational' is construed as stating that natural numbers k and l can be found such that $\sigma = \frac{k}{l}$. So again we have two sentences of the form 'a method x is known' and 'a method y is known' that, because of their form, cannot serve as the premise and the conclusion of the rule of the double negation elimination.

Thus, the upshot of this discussion is that the known examples advanced to derive the inadmissibility of rules or principles of classical logic from the neutrality thesis do not do their intended job. What they clearly show is rather banal, namely that the intuitionist, motivated by the neutrality thesis, interprets mathematical statements differently than his rival. Thus, we need to return to what emerged as the central problem, namely to the accusation that the classical mathematician misinterprets mathematical statements by holding to Platonism and that it is only Platonism that can justify his acceptance of bivalence.

3.2. *From Ontological Neutrality to the Repudiation of Bivalent Truth*

In essence, the classical mathematician's interpretation of mathematical statements is based on the assumption that each statement either holds or does not hold, and that this is so independently of whether or not it may be known whether or not it does so. It is just this assumption that the intuitionist challenges, his contention being that bivalence can only be justified by assuming some form of Platonism in respect to mathematical objects. This contention is motivated by a picture of some intuitive appeal, whose essential part is a distinction between statements that determinately either hold or not as the respective states of affairs obtain or not, and statements that cannot be so characterized, since the respective states of affairs are not completely present, or even cannot be completely present. According to the intuitionist, this incompleteness is brought about by mathematical infinity, statements of the second sort resulting from quantification over an infinite domain. In this view, our inability to complete an infinite sequence is not a mere medical impossibility, as Russell once said, but is the essential feature of infinity, which cannot be neglected in mathematical practice. This supposedly forces us to abandon a belief in properties of infinite totalities that obtain independently of our capacity to discover that they obtain. Thus, a property holds of an infinite totality only if this can be established on the grounds of what is known at a given stage about the generation of that totality, be it an initial segment of it, or some constraints on its future generation. Similarly, an infinitely proceeding sequence does not have a property only if this can be established at a given stage of its development, that is if a contradiction can be derived from the assumption that the sequence has that property together with the information about the generation of that sequence that is available at a given stage. In a similar vein, there is little sense in a sequence possessing the disjunctive property $A \vee B$ if this does not mean that at a given stage of the development of the sequence it is known how to prove that it has the property A, or it is known how to show that it has the property B. Granted these claims, the intuitionist demands abstention from the assertion that the sequence has the property or does not have it as long as neither of these disjuncts can be asserted. He claims, moreover, that the belief that the statement either holds or not, without any evidence for one or the other, misinterprets the true nature of infinity. Now, this intuitionistic ideology does not manifests itself in mathematical practice as long as it is known how to prove whether or not the statement holds, the controversy only being revealed in the approach to those mathematical statements relating to an infinite totality that remain undecided. Thus, in the case of first order arithmetic, we may limit our attention to undecided statements involving unbounded quantification.

Since our aim is to see how the classical mathematician may counter the intuitionistic objections in a way that cannot be accused of invoking the Platonist picture of mathematical objects, let us assume that he accepts the basic intuitionistic tenet that the statement holds only if it *can* be shown that it does. And, as an example of a controversial statement, let us take the hypothesis that there is the largest twin number. If so, the classical mathematician may counter his opponent's objection to bivalence in two ways. First, he may attempt to convey to his rival a suspicion that perhaps it is only a result of our ineptitude that from the already available information about the generation of the sequence of twin numbers we cannot derive whether or not the sequence terminates. Clearly, he should go further and claim that this may be so with every unsolved arithmetical problem. The intuitionist will obviously remain unconvinced, and ask his opponent for some grounds for being so optimistic. At this point the classical mathematician may resort to a second move. The natural numbers, whether thought of as a complete totality, or as an ever growing sequence, are introduced by means of the succession operation. Now, there are some truths concerning this operation, as for instance those that Peano attempted to capture in his axioms. And although the intuitionist may object to this or that axiomatization, he does not object to such truths as that each natural number has a unique successor, or that no natural number can have 0 as its successor or the principle of induction. Thus, the classical mathematician contends, it is plausible that these truths *determine* the answer to any arithmetical problem, although at a given stage of the development of mathematics some answers are not known. The intuitionist will remain undaunted by this contention and ask his opponent what he means by the answers being *determined* by basic arithmetical truths. If 'determination' means decidability in a given formal axiomatic system, then, as the Gödel results show, there are arithmetical statements that are not in this sense determined by the axioms. On the other hand, if determination is understood from the perspective of model theory, meaning that in each model of Peano arithmetic any closed formula is either true or false, the intuitionist would probably say that the metalinguistic proof of that theorem appeals to the excluded middle or other principles he objects to. Finally, the idea that basic truths of arithmetic determine an answer to each arithmetical problem may be understood as expressing a belief that for each statement which is undecidable in a given formal system, it will be possible to so improve the axiomatization that that particular statement will come out decidable in the resulting formal system. That, however, will not impress the intuitionist who will again accuse his opponent of being overly optimistic.

In this dual using pictures, with the intuitionist espousing the potential character of infinity and his opponent defending the view that the solution to any arithmetical problem may already be determined by the basic truths of arithmetic, we have reached a stalemate. However, the intuitionist has still another argument, which draws on his peculiar notion of the *infinitely proceeding sequence*. In general, one should view such a sequence as being more or less freely generated. At each stage of its generation an element is selected that accords with certain restrictions imposed at earlier stages, and further constraints on the future generation may be introduced. The crucial point of this concept is that both the selection of the elements of the sequence and the introduction of the restrictions are seen as essentially free. This is perhaps why it is better to conceive of them as being generated by an idealized subject than by some physical processes or automata. Now, taking this freedom seriously, there is little sense in the claim that the sequence has such and such a property, if this fact cannot be known on the basis of knowledge about the generation of a finite segment of it, that is, from what the elements of that segment and the imposed restrictions known at a given stage are. Similarly, it is hardly intelligible to suppose that a freely developing sequence does not have a given property, if this fact cannot be discovered on the grounds of how some initial segment of it has been generated. This strongly undermines the belief that generally, for a claim that a sequence has a certain property, one of two possibilities obtains: it either holds or not. If this is right, there is no reason to insist that bivalence holds for statements pertaining to choice sequences. In a similar vein, again taking seriously the freedom of generation, to assert that a sequence has property A or property B one should have a method of proving that it has one property or that it has the other, and that proof should be based on how some initial segment of the sequence has been generated—there is little sense in claiming that it has one property or the other if that cannot be known at a given stage. Similar remarks apply to assertions of quantified statements pertaining to the choice sequence. Since what emerges from these remarks is in essence the intuitionistic account of what constitutes a proof of the statement with a given principal operator, we have here quite a strong argument for intuitionism that draws on both the potential character of infinity and the freedom of generation of choice sequences. That the appeal to infinity is important is shown by the possible argument that, by (purported) parity of reasoning, the intuitionist should object to bivalence in the case of any statement pertaining to states of affairs whose obtaining is not determined, e.g., the statement 'Today John will drink tea at supper' if John has not made up his mind in this matter. The crucial difference is provided by there being a guarantee that either that statement or its negation will turn out to be true in the appropriate circumstances (supper time today), whereas, given the nature of the potential infinity, there is no guarantee that either an undecided statement pertaining to a choice sequence, or its negation will turn out to be true at some stage of the generation of that sequence.

How can the classical mathematician counter the above reasoning? First, concerning our example of the hypothesis of the largest twin number, he will notice that the generation of the sequence of twin numbers is not free in the above sense, as the requirement that its n-th element be the greater number of the n-th pair of twins does not leave room for any choice. He may also add that it is just the crux of the problem that it is not known whether that sequences continues infinitely, so how may one assume that it is an *infinitely proceeding sequence*? Nevertheless, though it may sound paradoxical, the development an infinitely proceeding sequence may break at a certain element, as no further elements satisfy the restrictions already imposed. (Brouwer 1952, p. 511) As to the first point, the intuitionist will agree that the generation of the sequence of twins is determinate (in his nomenclature it is a *law-like* sequence) but insist that the law-like sequence is a special case of the general concept of choice sequences, the other extreme being occupied by *lawless* sequences. Consequently, the logical operations figuring in the disputed claim should be construed as applying to the statement relating to the choice sequence. Thus, the assertion that there is the largest twin number or not is correct only if either it is known, on the basis of how the sequence of twin numbers is generated, that the sequence terminates, or it is known that it does not terminate. That means that at present the disjunction 'There is the largest twin number or there is not' cannot be asserted.

The other reaction of the classical mathematician may consist in repudiating the notion of the choice sequence. It is, however, implausible that choice sequences can be shown to be illegitimate on merely philosophical grounds, drawing for instance on their supposedly temporal or subjective character. Thus, as long as there are no intrinsic difficulties in the mathematics of choice sequences, there are no grounds to repudiate them. Moreover, the choice sequence and the related concept of spread were introduced by Brouwer to recover, along constructivist lines, the continuum of real numbers. The concept was thought of as an antidote to the classical mathematician's notion of the *arbitrary* sequence of natural numbers, as that latter notion clearly has Platonist underpinnings. The arbitrary sequence of natural numbers originates with the concept of infinite sequence whose progression is guaranteed by a law; then as the number of finitely statable laws in a given language is denumerable, one either assumes an idealization that there are, in some sense, undenumerably many laws of progression of which only some are statable in a given language, or, without taking recourse to laws straightforwardly, assumes that there are, in a Platonist sense, undenumerably many infinite sequences of natural numbers. All this means that a thorough discussion of choice sequences will lead to a controversy over the mathematical continuum and hierarchy of cardinalities, in which the Platonist underpinnings of classical mathematics manifest themselves very strongly.

The upshot of this discussion is that so long as the classical mathematician abstains from invoking Platonist assumption, he has no convincing arguments against the intuitionist's challenge to bivalence and to some forms of reasoning of classical mathematics. This in turn suggests that the intuitionist's observation that bivalence and the classical forms of reasoning must be based on mathematical Platonism is right. The above case for intuitionism draws on both the potential character of infinity and on the freedom of progression of choice sequences. However, as there is a popular view, reinforced by Heyting's writings, that the intuitionistic revision of mathematics somehow follows from assuming that mathematical constructions are mental, let us consider whether this supposition plays any role. We have already hinted that the idea that the choice sequence is generated by the subject, who imposes consequent constraints on its further generation, may have something to do with the problem of guaranteeing the freedom of the generation. Let us thus see if this hint is right. What lends support to this observation may be revealed in the following fictitious story. Suppose that while you are traveling in an aircraft, you spot ever more stars, and the continuity of these experiences makes you seriously entertain the hypothesis that if you traveled for ever, the sequence of the stars spotted would proceed infinitely. It happens that you are especially interested in white dwarves, and working on the hypothesis that the sequence of stars you would observe would continue *ad infinitum*, you begin to speculate about whether or not the sequence of observable white dwarves terminates. There is no information in the pattern in which white dwarves come and go that allows you to make a reasonable guess as to whether or not the sequence terminates. Now, the question is whether, given you share with the intuitionist the objection to the bivalence of mathematical statements, you should analogously protest against assuming bivalence for the statement 'The sequence of observable white dwarves terminates'. It seems that as long as you are not attracted by the view that stars are creations of your mind, your motivations for opposing bivalence, as applied to that statement, are considerably weakened. Moreover, if realism combines with the doctrine of an actual infinity, then the typical answer is that although it is not known which of the two possibilities obtain, the totality of observable dwarves is either infinite or not. However, if the emphasis is put on the potential character of infinity, things may turn out differently. There is probably still an inclination to say that given the way the subsequent white dwarves are brought forth, there is a determinate answer as to whether the sequence terminates or not. But this inclination seems to vanish once it is envisaged that the generation is truly free, that is either the objects are generated in a truly indeterministic physical process, or they are thought of as being created by a capricious agent who may impose some constraints on the subsequently generated objects, as the progression goes on. Thus, if these intuitions are any guide, the essential part of the case for intuitionism must be the conviction that the generation of sequences is free. Whether one tries to guarantee this freedom by appealing to their being mental constructions, or to truly indeterministic processes seems to be a secondary matter.

There is, however, a reasoning proposed by Michael Dummett that argues to the contrary, that the intuitionistic repudiation of bivalence and rejection of some classical forms of reasoning cannot be derived from the mental character of mathematical constructions with an implication that the traditional case for intuitionism, based on the potential infinity and freedom of generation of mathematical objects, cannot do its intended job. (EI, pp. 384–389; PBIL, pp. 231–243) The alternative advocated by Dummett is to present the controversy over intuitionism as concerning meaning of mathematical statements and in particular, the meanings of the logical connectives, where the decision about which meanings are correct, intuitionistic or classical, is to be arrived at through considerations of what the adequate representation of our knowledge of language is.

Thus, the question we need to address is whether someone who assumes with the intuitionists that natural numbers are creations of our minds may nevertheless consistently refuse to follow the intuitionistic revision of classical logic, as applied in reasoning concerning natural numbers. In the standard intuitionistic view, bivalence holds for a statement if an effective method is known for deciding whether or not that statement holds. Now, in first order arithmetic it is unbounded quantification that allows for the formation of statements for which the decision procedures are not known. Thus, the critical question is: given bivalent statements of the form $P(t)$, how can it be derived from the mental character of natural numbers that bivalence does not hold for $\forall_x P(x)$ or $\exists_x P(x)$, where for these latter statements the decision procedures are not known? Dummett's strategy consists in assuming an analogy between decidable statements that assign some properties to natural numbers (understood as creations of our minds) and statements about properties that fictional objects, say Shakespeare characters, may possess. He agrees that the analogy may be faulted by pointing out that in quantifying over fictional characters we are concerned with objects that *have been* created, while in the case of natural numbers, the quantification is over objects that *can* be constructed. This discrepancy, however, appears to be harmless, as long as it is maintained that the rules of construction of the natural numbers are determinate, and there does not seem to be any reason to oppose this view. Thus, translating our initial problem of quantification over the natural numbers, understood as products of our thought, into the issue of quantifying over fictional characters, we ask how the latter's ontological status forces us to abandon bivalence in the case of undecided statements resulting from the quantification over fictional characters. Now, the essential part of Dummett's strategy (but also, as we shall see, an instance of question begging) is to assume that the number of the fictional characters is finite. Given this assumption, the inclination to say that bivalence does not hold for the statement 'Each of Shakespeare's adult male character is shorter than 6 feet' must derive from the fact that the author did not decide how tall each of his characters is. That is, the opposition to the bivalence of the quantified statement $\forall_x P(x)$ stems from the failure of bivalence of the statements of the form $P(t)$. If this is any guide for how one may derive from the mental status of natural numbers the repudiation of bivalence for quantified arithmetical statements, bivalence must already fail for the arithmetical statements of which it has not actually been decided whether or not they hold, although the relevant decision procedures are known. Thus, by assuming the analogy between natural numbers, understood as mental creations, and fictional characters who form a finite totality, we have arrived at a position completely alien to the intuitionist. The result is, however, hardly surprising, since the analogy is demonstrably incorrect.

Clearly, in order to arrive at a better analogy, one should perhaps imagine a never-ending host of fictional characters. To be specific, let us imagine a never-ending soap opera that depicts in its subsequent episodes the lives of successive generations of some family. Obviously, we need to assume here that the profession of screenplay writer will never die out, so that the sequence of the fictional characters will progress *ad infinitum*. Now, there are certain properties, say color of hair, and being or not being the 'good guy', that the writers specifically assign to each of their characters. So, each of these properties, say being blond, being a good guy and the like, determinately either applies or not to each character. Does this mean that bivalence holds for such a quantified statement pertaining to these characters as 'Any character in this soap opera that has blond hair is a good guy'? If what is known from the already available episodes does not provide an example of a bad character with blond hair, one may argue against the bivalence of that statement along intuitionistic lines: As the sequence of the characters freely progresses, there is no sense in believing that any character of this soap opera that has blond hair is a good guy if this cannot be inferred from what is known about the episodes presented so far; similarly, there is no sense in believing that not every character with blond hair is good if this cannot be derived from the available information. Consequently, as long as there is no information testifying one way or the other, there is no ground for believing that the statement either holds or not.

Let us attempt to summarize what role the assumption of the mental status of mathematical constructions may have in the case for intuitionism. As we have seen, Dummett's argument to the effect that it does not play any role is fallacious, as it assumes a false analogy between natural numbers and a finite totality of fictional characters. This again points at the importance of the concept of potential infinity to the case for intuitionism. As to the mentalist assumption, apart from supporting such a concept of infinity and guaranteeing the freedom of generation of mathematical objects, it is intended to destroy the realist's distinction between a statement's being true and our ability to know that it is true. Nevertheless, given that there is some other way to blur this distinction, such as for instance the appeal to the free generation of choice sequences seems to provide, there is no need to invoke this assumption.

Be that as it may, with or without the mentalist thesis, we have here a plausible argument for intuitionism, whose premises are the potential character of infinity and the freedom of generation of choice sequences. Its first goal is to repudiate bivalence in reasoning about choice sequences. Secondly, it attempts to show that the acceptance of bivalence must presuppose Platonism. Finally, on the grounds of the conviction that mathematics should be philosophically neutral, it urges the revision of classical mathematics. Nevertheless, the weak point of this line of argumentation is the assessment of what is less philosophically neutral. Is it really so that the Platonist concept of arbitrary sequence of natural numbers is more philosophically loaded than the notion of choice sequence? We leave this question unanswered as we do not see any way to handle it.

4. THE SEMANTICAL ARGUMENT

We have seen that the intuitionistic challenge to bivalence calls, as Heyting's counterexamples make clear, for an interpretation of mathematical statements which would be quite different from the classical one. The first systematic investigations into what, according to intuitionism, mathematical statements should express is offered in Heyting's (1931) paper. As it aims to give an account of meaning for the language of mathematics, Heyting's ideas may be seen as predecessors of more contemporary arguments for intuitionism, whose starting point is an investigation into a correct model of meaning for mathematical sentences.

To begin with his reasoning, let us focus on the distinction he draws between sentence (*Aussage*) and statement (*Satz*), with the difference being that an asserted sentence is a statement. (Heyting 1931) How then is a sentence characterized? For Heyting, a sentence expresses a certain expectation, or, to use his other term, an intention. To take his example, the sentence 'C is a rational number' expresses the intention of finding two natural numbers k, l such that $C = k/l$. Intention, as he emphasizes, points to a possible result of our thinking, and not to a state of affairs that may transcend our ability to recognize it. (*ibid.*) The term 'sentence' is used to denote both a linguistic object expressing the intention and the intention itself. A statement, on the other hand, expresses the fulfillment of an intention. Thus, the statement 'C is rational' means that the natural numbers sought have been found. To point out that a sentence is asserted, or that it is a statement, Heyting uses Frege's assertion sign $\vdash$.

The concept of intention serves to introduce the logical constants or, as Heyting calls them, logical functions. These are characterized as operations on sentences (intentions) that permit us to build new sentences (intentions) from given sentences (intentions). To start with negation, a negated sentence, say $\neg p$, expresses a 'positive intention' which, on the one hand, attends to the intention p and, on the other, points to a contradiction. It is an intention of deducing a contradiction from the assumption that p. Thus, although in the original sentence, $\neg p$, there is no mention of a proof, the intention it expresses calls for finding a proof. (*ibid.*) Another logical function, disjunction, is explained as follows:

$p \vee q$ expresses an intention that is fulfilled only if at least one of the intentions: p, q is fulfilled.
(*ibid.*)

This characterization of negation and disjunction immediately entails that an instance of the excluded middle $p \vee \neg p$ is 'fulfilled' only if either a proof of p is known, or it is known how to deduce a contradiction from the assumption that p. Thus, the claim of the universal validity of the excluded middle, as it translates here into the claim that the excluded middle may be asserted for any mathematical sentence p, amounts to the conviction that any mathematical problem is solvable. Consequently, the universal validity of the excluded middle turns into a metamathematical issue.

Although Heyting does not go any further in the cited paper towards characterizing other logical connectives and quantifiers, one can easily supplement his account. Accordingly we may say that

$p \wedge q$ is an intention that is fulfilled only if both the intentions, p, q, are fulfilled;
$p \Rightarrow q$ expresses an intention of carrying out a construction that transforms anything that counts as fulfilling the intention p into a proof that fulfills the intention q.

Still further, on the assumption that we quantify over natural numbers, the quantifiers are explained as follows:

$\exists_x P(x)$ expresses an intention of finding such a natural number t that the intention $P(t)$ is fulfilled;
$\forall_x P(x)$ is an intention of finding a method that, if applied to any given natural number t, yields a proof that fulfills the intention $P(t)$.

Now, the troublesome aspect of this account is its very central feature, namely the appeal to intentions or expectations. The question arises how it can be guaranteed that various speakers ascribe the same intention to a given sentence. Without an account establishing the social character of intention, Heyting's explanation may invite accusations of being a version of the Humpty-Dumpty theory of meaning. This Lewis Caroll character once proclaimed: "When *I* use a word, (...) it means just what I choose it to mean—neither more nor less".[42] Clearly, whether or not such a charge can be leveled at Heyting's account depends on the understanding of his crucial terms: intention and the fulfilling of intention. In his account, while introducing these concepts, Heyting refers to Oskar Becker's book *Mathematische Existenz*. Becker, in turn, developed his teaching on meaning in accord with Husserl's theory from *Logische Untersuchungen* (LU). Husserl's basic idea is that expressions owe their meaning to intentional acts of consciousness, so that the meaning of an expression is determined by the intentional act with which it is uttered. It is just this idea that gives rise to Dummett's objection that in Husserl we still face a Humpty-Dumpty approach. (Dummett 1993, pp. 45–46) As there is little other evidence for how Heyting thought of meaning, apart from referring to the works of the phenomenologists mentioned, let us turn to their theory in the hope that it will shed light on Heyting's account.

Let us begin with how Husserl describes linguistic communication. As he has it, expressions serve as indications or signs of the thoughts of a speaker. (LU vol. I, p. 276) To put it in his nomenclature, through them a hearer is *intimated* with the speaker's sense-giving experiences, that is, experiences in which the uttered sounds receive meaning. This intimation may be understood in a narrow or wide sense. Given someone's utterance 'The dog is barking', it is only the speaker's act of judgment that is intimated in the narrow sense, whereas in the wide sense, other acts the hearer ascribes to the speakers, e.g., the speaker's acts of perception, are also intimated. (*ibid.*, p. 278) Understanding an intimation does not involve conceptual knowledge, it is not to be compared with judging or asserting. The hearer intuitively perceives the speaker to be a person expressing this or that. Moreover, given he perceives the speaker as manifesting some inner experiences, he also perceives to that extent the experiences themselves. Finally, as to mutual understanding, it "demands a certain correlation among the mental acts mutually unfolded in intimation and the receipt of such intimations, but not at all their exact resemblance." (*ibid.*)

When we turn to an expression having sense (*sense-informed expression*), Husserl draws the distinction between, on the one hand, the physical phenomenon of this or that particular sound or printed shape, and on the other, acts which give meaning to such a shape or a sound. These acts may possibly also give the expression its *intuitive fullness*, that is, permit the actualization of the relation between the expression and the object it refers to. Now, it is the central feature of expressions that they are directed to something objective, though what they are directed to may not exist. The feature of being directed-to is determined by the expression's meaning, whereas what the expression is directed to can either be actually present, that is intuited, or may appear as a mental image. If one of these obtains, Husserl says that the expression's relation to its object is actualized. However, it may happen as well that the object of the expression is neither intuited nor given as a mental image, and then the expression lacks acts that lend it intuitive fullness. What it surely has, is the *meaning intention that waits, so to speak, to be fulfilled.* (*ibid.*, p. 280) Husserl illustrates this by the way a name functions. As he says, whatever the circumstances, the name relates to the object in virtue of its meaning. As long as the object is neither intuited nor represented in mental imagery, there is only empty meaning-intention to that name. However, if a person establishes the bearer of that name, either by actually intuiting it, or representing it as a mental image, these formerly empty intentions become fulfilled, and a relation, of which the speaker is conscious, is established between the name and its bearer. (*ibid.*, pp. 280–281) Now, we need to take a closer look after Husserl at the acts already mentioned. They are of two sorts. On the one hand we have meaning-conferring acts or meaning intentions, that is "acts essential to the expression if it is to be an expression at all, i.e., a verbal sound infused with sense." (*ibid.*, p. 281) On the other hand, Husserl talks of meaning-fulfilling acts. These are claimed to "stand to it [the expression] in the logically basic relation of *fulfilling* (confirming, illustrating) and so actualizing its relation to its object" (*ibid.*) and are not essential to the expression as such. In any actual case, both sorts of acts are intermingled, so only in an abstract sense may it be said that the word is first uttered with a meaning-intention, and that this intention is then in turn united with the corresponding meaning fulfillment. (*ibid.*) Thus, as Husserl says, the word, as uttered to a hearer, serves the function of awakening a sense-conferring act in that person that points to what is intended, directing his attention to what may possibly constitute its fulfillment. (*ibid.*, p. 282) At this point a rather natural question arises about how acts of the first sort are related to those of the other or, in simpler terms, how the meaning intention determines what acts count as fulfilling. Let us first reflect on how we should think of this fulfilling. It should be kept in mind that we are investigating with Husserl the working of acts of consciousness. In line with Kantian tradition, the fulfilling acts should be viewed as operating, in accord with some categories of formation, on some data by which the object is presented. What results from these acts is not the object but rather the object's ideal correlate, that is the *fulfilling sense* of the expression. Now, Husserl claims that whenever the meaning-intention is fulfilled, then that ideal correlate of the object, that is, the manner in which the object is given, is the same as what the expression *means*. (*ibid.*, p. 290) In other words,

(...) [i]n this unity of coincidence between meaning and meaning fulfillment, the essence of the meaning fulfillment corresponds with, and is correlative, to the essence of meaning (...).
(*ibid.*)

Thus, the meaning intention is determinative of what the relevant meaning fulfillment is; moreover, the object of meaning intention is the same as what the fulfilling acts constitute. (*ibid.*, p. 291)

We have begun this plunge into Husserlian conceptions prompted by uneasiness about how it may be ascertained that a sentence, as described in Heyting's account, may have the same meaning for all competent speakers. Clearly, by the account discussed, acts conferring meaning as well as fulfilling acts may differ from speaker to speaker. Nevertheless, as Husserl stresses, the meaning of the assertion, as uttered on various occasions and by different people, remains the same. For instance, what is asserted in 'The sum of angles of a triangle equals two right angles' remains the same, no matter on what occasion and by whom it is uttered. This is contrasted with what is intimated, as the latter may consist of inner experiences of the speaker. Thus, what is asserted in the judgment, the content of the assertion, involves nothing subjective and, in contrast to inner experiences, neither arises nor passes away. (*ibid.*, p. 285) These observations lead to yet another distinction pertaining to what a linguistic object expresses: on the one hand it intimates sense-giving acts or perhaps sense-fulfilling acts of the speaker, so that the hearer becomes acquainted with that person's judgment and possibly his perceptions; on the other, the expression, properly speaking, expresses its content. (*ibid.*, p. 287) However, as a matter of fact, there are many uses of expressions that offer a *prima facie* counterexample to the ideally unchanging content introduced above. Husserl thus talks of changing meaning, or fluctuating expressions, and classifies them under the heading: *essentially subjective*. (*ibid.*, p. 321) Typically, these are expressions containing indexicals, or involving ambiguity or vagueness. However, their presence in the language does not, according to Husserl, undermine unchangeable meaning. As he says, the content of the so called subjective expression, as uttered on a particular occasion, is precisely the same as that of the appropriate objective expression. He insists that "each subjective expression is replaceable by an objective expression which will preserve the identity of each momentary meaning intention." (*ibid.*, p. 321) Accordingly, he maintains that changes of meaning of an expression, as uttered on varying occasions, are brought about by differences between acts of conferring meaning. (*ibid.*, p. 322)

How then shall we conceive of the identity of content? As Husserl has it, such identity is not a mere hypothesis, justifiable on some grounds or other: it is something immediately graspable, imposing itself with the utmost self-evidence. He thus states:

(...) [W]hat I mean by a sentence in question or grasp as its meaning, is the same thing, whether I think and exist or not, and whether or not there are *any* thinking persons or acts. (*ibid.*, pp. 329–330)

This holds true not only of the meaning of sentences, but also of the meanings of other types of expressions. Moreover, that identity holds true of such properties pertaining to meaning as being true, false, possible, impossible etc. To account for this identity, Husserl maintains that it must be thought of as an identity of *species*. The meaning of a given expression, as the species, contains the manifold of singulars identified with corresponding 'act-moments of meaning, the *meaning intentions*'. (*ibid.*, p. 330) Thus, the relation between meaning and various acts of conferring meaning, presentation and acts of presenting, judgment and act of judging is analogous to the relation that the species of redness has to a number of red slips of paper. Each of them, apart from having some individual aspects or other, like extention, shape etc., has its own individual redness, described by Husserl as an instance of that color-species. And this species is neither located in this or that particular slip of paper, nor does it exist somewhere else, and in particular not in our thoughts. (*ibid.*) The species is thought of as an ideal unit of meaning, not necessarily expressed by an actual expression. Even more, as Husserl states, there is

(...) no intrinsic connection between the ideal unities that in fact operate as meanings, and the signs they are tied, i.e. through which they become real in human mental life. (*ibid.*, p. 335)

The gap between such ideal meanings and expressed meanings is made even wider by Husserl's maintaining that not all ideal meanings are expressed meanings, and accordingly that there are many meanings that will not, or even cannot be expressed. Now, a question naturally arises about how the link between the sign and the ideal meaning is established, in cases in which there is such a link. Husserl does not seem to take a stand on this issue; nevertheless, there does not appear to be any grave problem with the view by which the expression the ideal meaning is related to is what we actually learn while mastering the language and that these relations are conveyed to new language users in the process of learning the language. With that addition, Husserl's avowed view that meaning is objective allows one to dismiss Dummett's objection that there is voluntarism in a particular person's act of conferring meaning on the expression. These acts should rather be conceived of as instantiating the ideal meaning in the uttered expression, in a fixed manner. Thus, the objection that competent speakers should inevitably differ in the meanings their respective acts confer on a given expression is groundless.

Nevertheless, it is clearly one thing to maintain that Heyting's semantical account, as based on Husserl's teaching, is not open to the objections discussed, and quite another to argue that it correctly prescribes the use of the logical functions. How can one produce a case showing that a negated or disjunctive sentence expresses exactly those intentions the phenomenologist ascribes to them? How may one convince his opponent that the fulfillment of the intention expressed by the disjunctive sentence must consist in either the fulfillment of one or the fulfillment of the other of its constituents? Thus, we confront here the task of deriving, on the ground of phenomenological analysis, the conclusion that Heyting's clauses relating to intentions hold and that any other understanding of the logical functions is incorrect. What emerges as the initial problem is the need to arrive at a notion of negation that authorizes a distinction between, on the one hand, the fact that the statement does not hold, and on the other, the fact that its negation holds. Clearly, this distinction is inherent in Heyting's account, as it permits us to distinguish between the sentence being not asserted, and the assertion of the negation of this sentence.

An endeavor to put negation under phenomenological scrutiny in the hope that it would support the intuitionistic reservations towards bivalence was undertaken in Oskar Becker's *Mathematische Existenz*, which appealed to Husserl's teaching on the fulfillment and frustration of the meaning intention, as propounded in the sixth *Logical Investigation*. Husserl's idea is that for those acts of meaning to which the distinction between the intention and its fulfillment is applicable, there is also a possibility contrary to fulfillment, namely frustration of the intention. Significantly, Husserl points out that frustration is not a merely negative fact, identifiable with simple not-fulfilling. What frustrates the intention should rather be thought of as a new fact, deserving characterization as a peculiar 'synthesis'. It is perhaps instructive to compare this with the fulfillment of the intention. The synthesis operative in the latter case is that of identification, since what the fulfilling act brings about, the *fulfilling sense*, is found to be identical with what the expression means. The synthesis peculiar to frustration of the intention, *distinction* as Husserl calls it, involves two features. First, "the object of the frustrating act appears *not the same as, distinct from* the object of the intending act." (LU, vol. II, p. 701) Secondly and equally importantly, for the object of one act to be different from that of the other, there must be a certain basis of agreement. There must be something that makes a frustrating act be the act *corresponding* to the intention. This may only derive from some fulfilling acts, the objects of which are identical with objects of the corresponding intentions. Thus, Husserl says:

> An intention can only be frustrated in conflict in so far as it forms a part of a wider intention whose completing part is fulfilled. We can therefore not talk of conflict in the case of simple, i.e. isolated acts. (*ibid.*, p. 702)

The doctrine can be amply illustrated by the following Husserl's example:

> If I think *A* to be *red*, when it shows itself to be 'in fact' green, an intention to red quarrels with an intention to green in this showing forth, i.e. in this application to intuition. Undeniably, however, this can only be the case because *A* has been identified in the two acts of signification and intuition. Were this not so, the intention would not relate to the intuition. The total intention points to an *A* which is red, and intuition reveals an *A* which is green. It is in the coincidence of meaning and intuition in their direction to an identical that the moments intended in the union with *A* in those two cases, come into conflict. (*ibid.*, p. 701)

Now, the question is whether, given that the concept of frustration of intention is to give rise to that of negation, there are in Husserl's account resources that enable us to establish a distinction between two sorts of negation, that is between not asserting the statement and asserting the negated statement. After Becker we may thus point to the notion of non-fulfilling of the intention, where non-fulfilling is meant as distinct from the frustration of the intention. Becker notices, however, that neither this nor a corresponding distinction is present in Husserl's doctrine; nevertheless, he proceeds to argue for it. (Becker 1927, p. 500) What lends support to the distinction between non-fulfilling and frustration of an intention is the inclination to regard the former as strictly negative, or having merely a 'privative' character. However, as long as something does not fulfill a given intention, there must be some agreement between a part, however small, of the intention and its fulfillment. Moreover, examples offered as putative cases of non-fulfilling have no such strictly negative character. To take Becker's example, let us considers the intention 'The book not lying on the table'. (*ibid.*, p. 501) Clearly, its frustration will consist in, first, locating or rather intuiting 'the' table as meant in the intention, and secondly, finding that 'the' book, as meant in the intention, is seen to be lying on the table. As to the non-fulfilling of that intention, Becker suggests that it happens if neither the book nor the table are seen in the vicinity of the person considered: neither 'the' book nor 'the' table, as meant in the intention, are intuited. Still, however, this is not a merely negative fact, since in order for both 'the' functions to indicate some particular objects, there must be something a person perceives, like the surroundings, say, a room, other pieces of furniture, and the like. So, in order for the fact of seeing neither the table nor the book not to fulfill the intention considered, there must be an agreement between the part of intention pertaining to the surroundings and the surroundings perceived. A similar moral follows from a more dramatic example involving people closed in a room and thus deprived of perceptions of the outside world. (*ibid.*) Taking the intention 'Birds not singing outside', as conferred on the appropriate utterance of one of the prisoners, it is hard to see that its non-fulfilling presupposes a positive agreement. Nevertheless, for that utterance to be meaningful on that occasion, or rather for the person to have the appropriate act of conferring meaning, there must be access to what the intention is directed to. And even in those unfortunate circumstances there is access to the object of intention, for instance, a part of the room leading to the closed door. Becker, moreover, maintains that for any intention there is access to the intended object. This is expressed by a principle he upholds, known as the *principle of access*, which states that for anything that may be an object there is, at least in principle, abstracting from technical difficulties, access. (*ibid.*, p. 502) Thus, even in cases similar to the envisioned situation of people deprived of perceptions of the outside world, the non-fulfilling of the intention must be based on the fulfilling of some parts thereof. Consequently, the distinction between non-fulfilling and frustration cannot be built on the differentiation between negative and positive fact, or between some agreement and no agreement at all.

It seems thus that to arrive at a substantial doctrine of not-fulfilling, as distinct from the frustration of intention, that could support intuitionistic views, we need to consider a variety of cases in which supposedly no frustration occurs, while the intention remains non-fulfilled. In a sense, for any person, there are many intentions that cannot immediately be fulfilled given the particular circumstances he finds himself in. Examples are provided by intentions pertaining to a region outside a person's vicinity, say 'snowing now in Moscow' considered by myself working now in my office in Cracow, or intentions requiring for their fulfillment the carrying out of elaborate calculations that, for contingent reasons, are not effected on a given occasion. There is probably another class of intentions, the fulfillment of which requires the person to find himself in circumstances to which access is physically or technically impossible, for instance intentions pertaining to spatially or temporally remote regions of the universe, or involving humanly unbearable physical conditions, like heat, radiation and the like. Finally, we may think of intentions such that although they are experienced as involving access to their objects, there is no guarantee that they can, in some strong sense, be either fulfilled or frustrated. Now, Becker's example of people deprived of perceptions of the outside world clearly falls in the first class, though perhaps it could be so reworked as to be classified in the second category. What we, however, need to achieve to get closer to the intuitionistic position is to conceive of the not-fulfilling of the intention as originating not from 'contingent' or 'technical' reasons. Otherwise, we will have to assume the third possibility, not-fulfilling, as distinct from the frustration of the fulfillment, even for intentions relating to trivial problems. There seems to be, however, a case in which the desired concept of non-fulfilling is credible: it is provided in mathematics by intentions relating to an infinite domain, where infinity is potential and the objects are thought of as somehow freely generated. But if this is so, we are back to the argument from choice sequences. Becker thus argues that given a choice sequence α, in generation of which the number 1 has not occurred so far, and considering the intention '1 occurring in α', there are three possible states of affairs (*Sachverhalten*) with which this intention is confronted:

1. the number 1 occurring in α;
2. the number 1 not occurring in α;
3. neither the state '1 occurring in α' nor the state '1 not occurring in α' being given. (*ibid.*, p. 504)

The three states of affairs give rise to the three respective statements and this, according to Becker, illustrates that in this case instead of 'tertium non datur', 'quartum non datur' holds. It is rather clear, however, that if one only asks why the above three clauses should replace the distinction between the statement being true and false, one is referred back to the doctrine of choice sequences. Thus, the argument, as it stands, adds little new to the reasoning based on the potential character of infinity and freedom of the generation of sequences, which we discussed in the last chapter.

It is equally important that even if we arrive at a satisfactory notion of not-fulfillment, we will still be facing the problem of deriving from it support for the intuitionistic revision of logic. Consider for instance the validity of the double negation elimination, taking to this end the familiar example of the rationality of σ. To put it in the present framework, the intention 'σ not being rational' is frustrated, as a contradiction follows from the supposition that it is; thus the sentence 'It is not the case that σ is not rational' can be asserted. The example requires us moreover to argue that the intention 'σ being rational' is neither fulfilled nor frustrated, that is, the state of affairs corresponding to this intention is that of 'neither σ being rational nor σ not being rational being given'. As the second part of this intention is unproblematic, the crucial question is why 'σ being rational' is not given? Why does the frustration of 'σ not being rational' not count as the fulfillment of 'σ being rational'? Obviously, Heyting's answer is that to assert the rationality of σ, we need to find integers k and l and show that σ equals k/l. (Heyting 1956, p. 17) But in order for this answer not to appear *ad hoc*, it should somehow be derived from the doctrine of intentions. That is, by analyzing intentions of the form 'x is a rational number', it should be shown that fulfillment of such an intention consists in nothing more nor less than finding integers k, l and showing that x equals k/l. It is hard to resist the observation, however, that the burden of such an argument far exceeds anything that can be achieved by phenomenological analysis. Likewise, it is hard to conceive of another justification of the proposed understanding of those intentions that, for instance, would draw on how sentences containing the logical connectives are actually thought of or used.

The upshot of this discussion is threefold. First, the objection to Heyting's semantic account, or rather the theory of Husserl and Becker that underlies it, according to which the appeal to the intentions makes it a Humpty-Dumpty account, is ill-founded. More specifically, there are no reasons to think that by this model people are thought of as ascribing meanings to expressions in some arbitrary way. As to the assessment of the semantical argument, it is to a great extent parasitic upon the reasoning that is built on the premises of the potential character of the infinity and legitimacy of choice sequences, as the latter likely provides grounds for the desired distinction between the frustration and the not-fulfilling of the intention. Finally, given that this distinction can be maintained, there is the crucial issue of how to arrive at the intuitionistic rejection of some laws of classical logic on the basis of the theory of intentions.

To end this chapter with a historical digression, it is easy to notice a similarity between later interpretations of the intuitionistic logical connectives and quantifiers that are given either in terms of proof or in terms of assertion, and the account framed with the help of intention. The basic idea of the proof-interpretation goes back to Brouwer's remarks on negation, which emphasize that in order to accept a negated proposition, say $\neg p$, one needs to derive a contradiction from the assumption that a proof of p has been effected. A systematic account in accord with this idea of Brouwer's was first propounded in Kolmogorov's (1932) paper, in which he characterized the connectives and quantifiers in terms of the notion of problem-solving, where 'problem' is a primitive notion. Assuming that letters a and b stand for problems, the explanation goes as follows:

$a \wedge b$ stands for the problem of solving both problems, a and b;
$a \vee b$ signifies the problem of solving at least one of the problems: a, b;
$a \Rightarrow b$ stands for the problem of deriving a solution of b from a solution of a;
$\neg a$ signifies the problem of deriving a contradiction from the assumption that a solution of a is given;
$\forall_x a(x)$ is the problem of finding a general method that yields a solution of the problem a for any given t;
$\exists_x a(x)$ is the problem of finding a certain t and solving for it the problem a.

(Kolmogorov 1932, pp. 60–61)

As a virtue of the above account, one should consider its emphasis on the intersubjective meaning of the logical connectives. As Kolmogorov remarks, in the case of everyday problems, their solutions are often purely subjective facts that may be of no interest to other members of a community. Also, opinions may strongly diverge as to what counts as a solution of a problem. But mathematical and logical problems, on the contrary, possess a special feature: their solutions have universal validity. Even more, a mathematical or logical problem is solved only if its solution is presented in a way that allows its correctness to be universally recognized. Moreover, the recognition of its correctness is necessary for acknowledgment of the solution. (*ibid.*, p. 61) We may add further that possible differences of opinion among mathematical schools concerning what counts for a solution of a problem do not present any persistent difficulty on this account, since one can always stipulate in the description of a problem what means of proof are admissible. Hence, despite ideological differences, a problem 'means the same', no matter what philosophical orientations mathematicians may represent.

Kolmogorov's interpretation was approvingly quoted in Heyting's (1934) paper; nevertheless in his later writings Heyting chooses still another, though closely related, interpretation. He invokes the notion of assertion to serve the objective of explaining logical connectives and quantifiers. To give some examples, his account requires that

$p \wedge q$ can be asserted if and only if both p and q can be asserted;

$p \vee q$ can be asserted if and only if at least one of the propositions p and q can be asserted;

$p \Rightarrow q$ can be asserted if and only if we possess a construction t, which joined to any construction proving p (supposing that the latter be effected), would automatically effect a construction proving q. (Heyting 1956, pp. 98–9)

Now, since the assertion of a statement is legitimized only by knowledge of the appropriate proof or construction, the assertion-interpretation naturally boils down to characterizing the connectives and quantifiers in terms of proofs, which is now an accepted standard. Again, to look at some examples:

a is a proof of $\varphi \wedge \psi$ iff a is a pair $\{a_1, a_2\}$ such that a_1 is a proof of φ and a_2 is a proof of ψ;

a is a proof of $\varphi \vee \psi$ iff a is a pair $\{a_1, a_2\}$ such that $a_1 = 0$ and a_2 is a proof of φ or $a_1 = 1$ and a_2 is a proof of ψ;

a is a proof of $\varphi \to \psi$ iff a is a construction that converts each proof b of φ into a proof $a(b)$ of ψ. (van Dalen 1986, p. 231)

Granted these intuitive stipulations, two tasks emerge: first of erecting a formal semantics that will yield the meanings of the logical constants as stipulated above, and second, of finding some reasons for preferring this account to that of classical logic, and consequently, for repudiating the classical meanings of the logical connectives. As one should expect, Heyting's rationale for objecting to the classical interpretation of the logical operations draws on the criticism of the concept of truth. Thus, while explaining the intuitionistic implication, he complains that the traditional definition, i.e., $p \Rightarrow q$ is false if and only if p is true and q is false, is meaningless for the intuitionist if the logical values of p and q are unknown. (1934, p. 390) This suggests that for him any meaningful talk of logical values presupposes the ability to recognize what logical values sentences have. Why, however, should one dispose of the notion of objective truth (and respectively, falsity), a concept that permits us meaningfully to talk of the logical value of the sentence despite our not recognizing, or perhaps even not being able to recognize, whether it is true or false? The answer should be sought in the neutrality argument, as support for the repudiation of the objective concept of truth can be traced back to a postulate that mathematical investigations should be carried out without invoking the fiction of a realm of mathematical states of affairs that make some propositions true and the others false. This again supports the conclusion we have already drawn that the fate of the semantical reasoning depends on the success of the neutrality argument.

4.1. *A Note on Heyting's Views on Formalization and Logic*

Heyting's best known achievement is the axiomatization of some parts of intuitionistic mathematics and the formalization of intuitionistic logic. With these results, published between 1925 and 1930, it is natural that discussions over intuitionism began to concentrate more on the interpretation of the logical connectives and quantifiers than on strictly mathematical matters. Remembering Brouwer's sometimes hostile attitude to logic and formalization, it is somewhat surprising to find, as van Stigt's (1990) book documents, that the father of intuitionism was encouraging his student to take an interest in that type of research. Despite the encouragement and Brouwer's lively interest in the debate about intuitionistic logic, his evaluation of formal studies of mathematics remained unchanged: it is unproductive because it concentrates on studying the language of mathematics, and not mathematics itself as developed in processes of mental construction. This stance of Brouwer's was shared by Heyting in his papers published in the 20's and early 30's. Even more radically, he stresses there that mathematical construction 'transcends' any formal systems, as no system of formulae and rules may contain all the possibilities inherent in human mathematical thinking. (1934, p. 388) However, the move towards formalization brought about a welcome turn towards the use of symbolism in intuitionistic mathematics. Brouwer's *penchant* for avoiding, whenever possible, mathematical symbols was replaced by a more liberal attitude towards symbolic notation. Nevertheless, despite the possible differences in assessing the role of logic, both mathematicians agree that there is no essential difference between formal and ordinary languages: according to them both sorts of languages share the same limitations. In particular, formal language can neither guarantee the certainty of mathematics, nor exclude misunderstanding between mathematicians. In spite of this, Heyting expresses a conviction, rather alien to his master, that symbolism proves an effective medium for understanding notions of intuitionistic mathematics and conveying its proofs to readers. As he puts it,

Though mathematical constructions are originally mental operations, symbolization is necessary to a certain extent, for practical reasons. However, in a constructive theory the symbolic logic will be considered to express mathematical constructions, so that every formula will admit an immediate and unique interpretation. (...) It is sometimes clarifying to formalize a chapter of constructive mathematics, but it is never necessary. (Heyting 1959, p. 70)

Heyting's preoccupation with logic makes one expect that his opinion as to what role the principles of logic play in mathematics will be more liberal than his master's; in particular, it might be expected that some principles will be acknowledged as a foundation for intuitionism. This expectation is, however, rather mistaken. Heyting shares with Brouwer a twofold opinion as to what logic is. Accordingly, principles of logic are either the most general mathematical theorems about 'possibilities of embedding' one mathematical system into another, or laws of applied mathematics that give accounts of admissible transformations of sentences concerning mathematics. (Heyting 1934, pp. 389–390) On the other hand, Brouwer's early extravagance of equating logic with ethnography is never mentioned in Heyting's papers. Instead, he supplements the two-fold Brouwerian position with a third interpretation of logic, according to which laws of logic are 'about the world', or, in other words, ontological theorems. To clarify these matters a bit, let us take a look at one of his examples. (Heyting 1974, p. 86) Consider:

(1) Socrates is a man.
(2) Every man is mortal.
---
(3) Socrates is mortal.

The first interpretation is that the rule applied above is a 'rule of language' of the following form:

(1) A is a B
(2) Every B is C
---
(3) A is C

The rule, on this interpretation, states that whenever a person agrees with statements (1) and (2), he is also expected to agree with (3). On the second interpretation, the rule states something 'about the world'. Given that the world is such that it makes (1) and (2) true, it also renders (3) true. Finally, one may read this as a mathematical theorem: 'If the entity A belongs to the species B, and B is a part of the species C, then A belongs to C'. In this way we obtain three interpretations of what logic is. Heyting stresses that on the first and second account logic is a sort of applied mathematics while the third interpretation translates logical principles into theorems of set theory, a rather abstract part of mathematics. Consequently, he concludes that logic cannot, on any of its interpretations, play any role in the foundations of mathematics. However it is interpreted, it inevitably presupposes mathematics. (*ibid.*)

The debate over intuitionist logic, in which Heyting played a vital role, brought about in the long run a significant change in the intuitionistic program, a development hardly agreeable to both Heyting and Brouwer. Most importantly, efforts aimed at justifying intuitionism began focusing almost exclusively on logical questions. Thus, the problem of replacing classical connectives and quantifiers by the intuitionistic ones became crucial. Parallel to this, Brouwer's followers set about the more pragmatic task of finding formal, axiomatic theories of species, choice sequences, or spreads, largely leaving aside the enterprise of finding reasons why these theories should be preferred to their classical counterparts. Still another influence of Heyting's studies of intuitionistic logic that one could perhaps mention is the bearing they may have had on the later views of Brouwer. In one of his last papers (1955) one finds him expressing a rather mild view of logic, even praise of it. While finishing his talk on the impact of intuitionism on classical logic, he thus concedes:

> Not only as a formal image of the technique of common-sensical thinking has [classical algebra of logic] reached a high degree of perfection, but also in itself, as an edifice of thought, it is a thing of exceptional harmony and beauty. (Brouwer 1955, p. 554)

5. INTERSUBJECTIVITY IN HEYTING'S CONCEPTION

The many explicit remarks praising communication between mathematicians that appear in Heyting's papers suggest that he saw the necessity of improving Brouwer's conception of mathematics as a *languageless* mental creativity so that the danger of the lack of intersubjectivity could be avoided. For instance, while describing the intuitionistic position in his (1958b) paper, he states these two principles:

> [Intuitionistic mathematics] cannot do without linguistic expression in words or signs, but the latter serve only as support for memory and for communication; formulae without the background of a mental construction have no place in it; (...) Understanding between mathematicians is essential. (*ibid.*, p. 103)

To ensure the possibility of communication Heyting postulates a similarity between people's thoughts. As he puts it:

> (...) My mathematical thoughts belong to my individual intellectual life and are confined to my personal mind, as is the case for other thoughts as well. We are generally convinced that other people have thoughts analogous to our own and that they can understand us when we express our thoughts in words (...). (Heyting 1956, p. 8)

Among analogous thoughts, constructions of natural numbers purportedly have a distinguished role, as he believes that every person, including schoolchildren,

> (...) knows how to build mentally the natural numbers, and also knows what it means that their sequence can be made to proceed indefinitely. (1958b, p. 102)

Elsewhere he adds, moreover, that

> (...) [t]here is strong evidence for the hypothesis that the construction of small natural numbers is the same for all [people] (1974, pp. 88–9)

but concedes that more complicated structures can be thought of differently, so misunderstandings are not excluded. One may nevertheless ask what it is that guarantees that mathematicians are able to carry out the same (or analogous) mental constructions. Contrary to Brouwer's teaching entailing the intuition of two-ity as a defining feature of mind, Heyting does not deploy any similar speculative reply. For him, it is simply a psychological fact that people's thoughts are 'analogous', a fact that he does not explain any further. More worryingly, he does not distinguish in mathematical thought between the subjective element (called 'emotional content' in Brouwer's terminology) and the strictly mathematical, though mental, system. Thus, we have in Heyting the willingness, or the perception of the need, to satisfy the mentalist condition of intersubjectivity, though its satisfaction is maintained in a rather *ad hoc* manner.

The way the problem of communication is set suggests that Heyting stands on the ground of a mentalist, Lockean picture of the functioning of language. Thus, mathematical thought belongs to a person's individual intellectual life and consequently is confined to the mind of that particular person. Mathematical thought may nevertheless be expressed in words, and somehow communicated to others, as language is conceived of as meeting the task of influencing the thoughts and actions of others. (*ibid*.) Consequently, lecturing on mathematics and writing mathematical papers or books aim to suggest mathematical constructions to other people. Similarly, when a mathematician puts down some notes to aid his memory, he is doing so for his 'future self' that plays the role of the other person. (*ibid*.) Given these views, our question is whether they conflict with the Wittgensteinian condition of communicability. We have argued in chapter 2 section 4.3 that the assumption that mathematical objects are mental constructions does not, on its own, lead to the view that people cannot communicate about them; it is rather the concept of the private naming of such constructions that may lead to this consequence. Thus, to decide whether Heyting's somewhat metaphorical claims give rise to the charge of violating intersubjectivity, we need to consider what concept of naming Heyting accepts: are mathematical mental constructions given names by 'private' ostensive definition. However, the problem is that Heyting's writings contain only very scanty material that could cast light on what views of the naming of mathematical objects he holds. We are thus left with the rather speculative task of supplementing his account of linguistic communication with an explication of whether and how mathematical structures receive their names. As such supplementing runs the risk of ascribing to Heyting views that he most probably never actually held, we should rather think of it as addressing questions like what tenets of his philosophy could force him into accepting a model of private naming or, the other way round, whether his doctrine contains some claims that prevent him from accepting a social model of naming and meaning.

To start with, there is a part of his doctrine that makes one think that his views do not entail the troublesome concept of naming: namely the psychological account of the origination of mathematical concepts. This account stresses that it is a real, flesh-and-blood person who learns to distinguish 'wholes' in the field of perception, then starts thinking in terms of those wholes and finally, after mastering the art of abstracting from the sensuous contents of perceptions, arrives at the concept of one. A further psychic faculty, of focusing one's attention on still another perception and repeating the process of abstraction, makes possible, according to Heyting, the formation of the concepts of arbitrary number and indefinitely proceeding sequence.

Now, since Heyting never suggests that these processes do not take place in social circumstances, it is easy to supplement these semi-psychological ideas concerning the origination of mathematical concepts with a social account of mastering mathematical expression and, perhaps, mathematical concepts as well. Without going into the details of such an account, we may consider a child who is being taught words for small natural numbers playing, say, at blocks. To put this in Heyting's terms, since the child already sees the blocks as different entities, there is nothing to prevent it from mastering expressions like 'two blocks', 'three blocks', 'five dolls'. In parallel, it is taught to memorize the litany of number words, so that, if suitably prompted, it will recite 'one, two, three, four, five'. Obviously, the older the child, the longer the sequence of number-words it can recite, and the fewer mistakes it makes. At some stage these two abilities, of telling the number of things and saying the sequence of number-words, combine to yield the capacity of using numbers quite abstractly, that is without reference to objects seen on a given occasion. Reverting back to Heyting's nomenclature, ability to abstract from blocks, dolls, apples, and other perceptible things, is the precondition for mastering the abstract use of numbers.

Now, if we think of mathematical concepts or constructions as our thoughts or perhaps parts thereof, how should we conceive of what the social process of the learning of mathematical expressions accomplishes? Should we view it as resulting in establishing a link between the expression, as functioning in the common language, and someone's mental concept? If we do not assume any such link, then it is hard to say what role the appeal to mental concepts plays. That is, the child's ability to arrive at 'pure wholes' in his mental processes would be indispensable for mastering the number-words; but then for involvement in mathematical tasks, or communicating with others, it would be otiose to suppose that the expressions uttered were *linked* to appropriate mental objects. On the contrary, abilities to solve mathematical tasks and communication of the results would be explained exclusively in terms of the application of *rules* of handling various mathematical symbols and expressions, rules that are acquired from the community. Clearly, the qualification would be added that to get this process started, there must be a certain regularity in the human psychological setup, as, for instance, assumed in Heyting's idea of a common capacity to discern entities. On the other hand, if one opts for there being a link between words of a common language and mathematical concepts of a mental character, a question arises about the nature of this link. Are we forced to say that the expression *refers* to a mental construction? What comes to mind is Wittgenstein's distinction between the word referring to the sensation and the word replacing the natural, primitive manifestations of the sensation, the paradigmatic example of the latter being how the word 'pain' replaces cries and other pain behavior (*Philosophical Investigation* I, §244). This distinction, however, offers only little help, if any, since there is little credibility to the idea of mathematical concepts having natural and primitive manifestations. Perhaps it could be possible to frame some other concept of the link, as distinct from referring, but as it is hard to imagine what would it be, it seems that we are left with referring. What, however, is troublesome with viewing the words as referring to mental concepts, given that the reference is established under the control, or influence, of the linguistic community? The problem is exactly this: how far the supposed control extends or, similarly, what it can achieve, as it is ultimately a particular person who *decides* what in his mental inventory is to be denoted by the word he hears in other people's mouths. Since we have discussed this issue while investigating intersubjectivity on Brouwer's conception of mathematics, we only repeat our main points here. The first trouble is how to conceive of that decision. If it is conscious, or at least can be brought to consciousness, then the picture is unrealistic; if not, how can we at all talk of a *decision*? Secondly, if the words referred to mental constructions, each of us would talk only of his constructions. By assuming that we have perfectly analogous constructions, we still may guarantee that we talk about qualitatively the same things, though numerically different, but, more worryingly, it appears that the truth of a statement would be a property relative to the person who asserted it. Now, if we somehow manage to circumvent these problems, should we assume that the naming is correctable under the influence of other language-users, or not? If not, we have here only a camouflaged account of private naming with the implications that the issue of correctly remembering the name cannot be addressed and mathematical results are not communicable, at least by means of language. On the other hand, if it is correctable, this feature is wholly supervenient on the person's use of the word being in agreement with the word's use in the linguistic community, which makes otiose the appeal to mental constructions as something the words refer to.

Thus, two ways of supplementing Heyting's account emerge from this discussion which are free of the undesirable consequences. The first is to deny that mathematical expressions refer to mathematical constructions and to maintain that this inventory of mental constructions is needed only in so far as it provides a necessary condition for the start of language-learning. The other option also abandons the idea that words refer to mental constructions, but optimistically counts on finding some other way of conceiving the link between the two. This means, however, that at present we have only one serious proposal as to how to supplement Heyting's conception of mathematics in order that the Wittgensteinian condition of intersubjectivity not be violated. Naturally, the question is whether this proposal does not conflict with Heyting's position.

Now, it may be objected that the proposal runs counter to Heyting's views, as linguistic communication consists, according to him, in influencing the thoughts of others, or invoking the appropriate mathematical thoughts in their minds. However, this view does not commit its proponent to maintaining that mathematical expressions refer to thoughts or parts thereof. As a matter of fact, there are many ways of using language that can reasonably be described as influencing the thoughts or feelings of others, where the words' referring role is negligible. Thus, the crucial question is whether Heyting had ever actually required mathematical expressions to refer to mental concepts. As far as we know, this must be answered in the negative. Besides, one might also argue for a stronger position, namely, that Heyting's opinion on the role of mental concepts in the functioning of the language of a mathematical discourse not only is not in conflict with the view proposed here, but that there is an affinity between the two. An argument of this sort will draw on the evolution of Heyting's views that gradually put less emphasis on the mental character of mathematical objects and more on symbolization. Thus, in his papers of the 1950's one finds a concession that symbolization is necessary for practical reasons as a means of memorizing and communication.[43] This stance is reinforced in (1956, p. 15), where while considering in what sense a large natural number symbolized by a long series of digits can be given to the mind, he concedes that mathematicians deal with symbols, and a demand of constructivity should be seen, in such cases, as a somewhat idealized possibility of counting until the number is reached. At this point one may wonder what the series of digits refers to. It is somehow counterintuitive, though not patently wrong, to claim that the denotation is a construction that eventually forms in the mind of the person who carried out the appropriately long process of counting, with the added caveat that the carrying out of this construction may not be humanly possible. Nevertheless, the issue fades into irrelevance once we learn from Heyting that the meaningfulness of a long series of digits has much to do with the fact that mathematicians deal with symbols. One understands the series of symbols since one knows what it is to count up to that number, that is how to arrive at that long row of digits by observing the relevant rules for handling these symbols, where these rules are learned in the community. Similarly, by someone's utterance

'The number 74591983974003013747927 is prime'

communication is achieved not because the hearer miraculously forms a construction to which the series refers, but since he knows various rules for operating arithmetical symbols, as these rules function in his community. Some of these rules, moreover, allow him to decide whether or not a given property holds of that number, for instance whether it is prime. Thus, the argument will try to establish that once it is conceded that the meaningfulness of mathematical expressions derives from socially established practices of observing rules regulating operations on relevant symbols, there is no need to assume that mental constructions provide denotations.[44]

To sum up these remarks, one cannot object to Heyting's conception of intuitionism on the grounds that it violates the mentalist condition of intersubjectivity. First, he seeks the basis for successful communication among mathematicians in 'analogous thoughts', despite the fact that these are introduced in a somewhat *ad hoc* manner. Secondly, as to the Wittgensteinian condition, no part of Heyting's doctrine leads to its violation. Moreover, his psychological explanation of the origination of mathematical concepts together with his acknowledgment of linguistic means in mathematical practice, allow his conception to be supplemented by a social model of the learning of mathematical expressions.

6. RESUME OF HEYTING'S ARGUMENTS

For anyone who looks with suspicion or dislike at Brouwer's metaphysical extravagance, Heyting's case for intuitionism appears a sober and welcome development. He abandons almost entirely the spurious machinery of Brouwer's philosophical system: consciousness, mind, causal and temporal attentions, intuition of two-ity, linguistic attention. Although many Brouwerian terms remain, they receive a different meaning. The 'basis for mathematics' is sought in psychological faculties and intuitionism is presented as a metaphysically neutral position. Also, the extreme mentalism of the founder of intuitionism is replaced by a more liberal stance towards symbolization and formalization.

Turning to the question of the intersubjectivity of mathematics according to Heyting's conception, his remarks show that he was aware of the problem, although on the other hand he seems to believe that intersubjectivity can be secured by the mere satisfaction of the mentalist condition, as we call it. The condition is met in a somewhat *ad hoc* way by claiming that people have analogous thoughts, whereas the nature of this similarity and its origination are left unexplained. As to the Wittgensteinian condition, that is, one requiring that people be able to acquire a common language serving for communication, it is met once Heyting's account of the origination of mathematical (mental) concepts is supplemented with a social account of the learning of mathematical expressions, the further demand being that mathematical expressions do not refer. We have argued, however, that Heyting's doctrine is neutral as to whether mathematical expressions refer. In general, Heyting's later emphasis on the need for symbolization and collaboration among mathematicians has the consequence of making the satisfaction of the Wittgensteinian condition much easier than is the case in Brouwer's philosophy.

Our second problem pertains to semantic theory and its relation to the concept of truth. In contrast to Brouwer, Heyting has such a theory, and it is based on the teachings of the phenomenologists: Husserl and Becker. Husserl's doctrine of intention and its fulfillment serves to explain the meaning of mathematical sentences, in particular those involving the logical constants. However, turning to the question we posed in the Introduction, that is, whether the semantic theory provides reasons for repudiating bivalent truth, one first notices a discrepancy between Husserl's and Becker's views of the non-fulfilling of the intention, the issue having a direct bearing on the concept of negation. Following Becker, we indeed have three possible states of affairs, corresponding roughly to assertion of the sentence, assertion of its negation, and abstaining from either assertion, but on its own this trichotomy does not give any reasons for abandoning bivalent truth. The argument for the latter, and in fact a justification of the trichotomy, turns on the concept of the choice sequence, the premises being the potential character of infinity and freedom of generation of those sequences. Thus, to argue for the correctness of this semantical account, one needs first to have convincing grounds for repudiating bivalent truth in reasoning pertaining to choice sequences. Therefore, in the case of Heyting's philosophy, it is not the semantic theory that is seen as leading to a revision of the concept of truth, but the other way round.

This brings us to the problem of assessing Heyting's arguments for intuitionism. We have fished out of Heyting's papers three types of such arguments: psychological, neutrality, and semantical, the weakest being the first, the strongest the second, and the third being parasitic on the second. The psychological reasoning has its source in Heyting's taking the Brouwerian intuition of two-ity for a faculty comparable with seeing some statements or mathematical constructions as evident, the implication being that constructions have an evident character, as they can be seen intuitively. This however can hardly convince the classical mathematician who surely considers his proofs more evident than intuitionistic. More importantly, Griss's schism over negation shows that even in the intuitionistic camp there is no agreement as to what is evident. Leaving aside the semantic argument since it adds little to the matter, let us now turn to the ontological reasoning, which provides, in our opinion, the strongest debating position. Its crucial ingredient is the conviction that mathematics should be neutral in respect to metaphysical matters: it should not work on the premise that mathematical objects exist timelessly and independently of the mathematical activities of human subjects. This provides motivation for holding that the only legitimate concept of infinity is that of potential infinity. In addition, it is assumed that the choice sequence is a legitimate object and, moreover, the concept of choice sequence is the general concept of infinite sequences, a special case being the law-like sequence. Now, the strategy of the reasoning relies on inveighing against the unqualified applicability of truth-values to statements pertaining to choice sequences or, more precisely, against divorcing the truth of such a statement from the derivability of that statement from the information about the generation of the sequence. The argument appeals to the freedom of the generation of choice sequences; hence supposedly the need to assume that they are mental, that is, generated by a free subject rather than a physical process. A further part of the strategy consists in maintaining that the logical constants occurring in those statements cannot be framed in terms of bivalent truths. In the final step, it is maintained that the conclusions carry over to sequences familiar to the classical mathematician, that is law-like sequences, as they are considered a special case of the general notion of choice sequence. The reasoning, as we see it, ends in a sort of stalemate. On the one hand, the uncountably many infinite sequences of natural numbers that the classical mathematician works with clearly presuppose Platonism, but on the other hand, the objects the intuitionist assumes look rather odd to mathematicians, though perhaps from the metaphysician's perspective they are less demanding than Platonist entities. Worse still, various intuitions concerning freedom, time, potential infinity, and others are far from clear, and sometimes even seem to be in conflict. Where then shall one turn to resolve the dilemma? It seems that we have reached a stage at which the problem should be handed over to mathematicians, hoping that the relations between classical mathematics and the mathematics built on the notion of choice sequence will shed more light on the issue than investigations of unclear and sometimes conflicting intuitions.

4
DUMMETT'S CASE FOR INTUITIONISM

1. DUMMETT'S PROGRAM: AN OVERVIEW

At the time when Heyting was announcing peaceful collaboration between competing schools in the philosophy of mathematics and other intuitionists were adopting a more pragmatic program of formalizing various chapters of intuitionistic mathematics, another reasoning that advocated the necessity of revising classical mathematics in favor of intuitionism was crystallizing in the publications of Michael Dummett. It was not only new and different from those of Brouwer and Heyting, but, as it started with considerations of what a correct account of meaning should be, it marked a significant breach with the already established intuitionistic tradition. Although a part of what Brouwer and Heyting wrote on intuitionism can clearly be viewed as expressions of their views on meaning, their arguments ultimately turn on foundational issues like the nature of infinity or the concept of set rather than on reflections on meaning.

To place Dummett's reasoning among other arguments for intuitionism, it should be noted that his argumentation is limited to the problem of revising logic, and not the whole of mathematics. One reason for this limitation comes from his conviction that the most fundamental feature of intuitionistic mathematics is its underlying logic. (PBIL, p. 215) Accordingly, he is preoccupied not with a general query about whether the practice of classical mathematics is so essentially flawed that it needs to be abandoned and replaced by intuitionism, but with possible reasons for replacing, in reasoning with mathematical statements, the canons of classical logic with those of intuitionistic logic. By 'logic' in this context he understands sentential and first-order predicate logic. Dummett's reasoning and particularly his examples are moreover limited to first-order arithmetic. It appears that this choice is somehow unfortunate because intuitionistic and classical arithmetics are, after all, translatable. Thus, a serious controversy between the two schools reveals itself in more advanced theories, like set theory or theory of real numbers. All this suggests that it is not clear whether the possible success of Dummett's argument for revising the forms of reasoning of classical logic applied in a mathematical discourse will lead to the intuitionistic revision of *mathematics*—whether, for instance, classical set-theory or theory of real numbers will yield to their intuitionistic rivals. Thus we may see Dummett's project as assuming a division of labor, his part being to provide arguments for the revision of logic whereas the task of defending intuitionistic mathematics against possible objections of inconsistency, lack of clarity and the like are left to others. However, only if both components of this program are carried out will we have an argument for preferring intuitionistic over classical mathematics.

The above remarks are not intended as criticism of Dummett's starting point. They attempt to make clear the essential differences between, on the one hand, Brouwer's and Heyting's approaches and, on the other, that of Dummett. Examining arguments of the early intuitionists, we came across such fundamental themes as: Does mathematics need a foundation and, if so, what should it be? What is the relation between mathematics and logic? What are mathematical objects and how may they be introduced? We have also seen that along these fundamental lines there are no clear-cut arguments for the intuitionistic program of revising mathematics. Accordingly, as long as classical mathematics is not shown to be inconsistent, the choice between it and intuitionism is rather a matter of taste than of conclusive argumentation. Thus, concentrating on an important but by no means crucial tenet of intuitionism, one may hope that this dispute in the philosophy of mathematics will become more 'manageable' and offer at least some chance of being resolved. Still another ground for being optimistic in this respect is that differences between intuitionistic and classical logics can be stated at a fairly elementary level, at least when compared with differences between chapters of intuitionistic and classical mathematics. But granted that it is made clear how the two schools diverge in assessing the validity of the laws of logic and rules of inference, the crucial question emerges as how to decide which logic prescribes the correct forms of reasoning with mathematical statements. Now, an important observation in this respect is that a controversy over the grounds on which a sentence may be asserted, or what conclusions follow from a given sentence, may result from a disagreement over *meaning*. Accordingly, one may believe that what is at stake in the controversy over intuitionism vs. classical mathematics is the meanings of mathematical expressions and, in particular, of the logical constants. If the controversy concerned some more advanced mathematical theory, then these two problems, the meanings of logical constants and the meanings of mathematical expressions, would be entangled; for instance a dispute over a correct theory of real numbers would of necessity draw on the meanings of the logical constants as well as on the interpretation of such fundamental notions as infinity, convergence of sequences of rational numbers, ordering relations and real number. However, given that there is a theory whose concepts are not the subject of a heated dispute between the schools, which is exactly the case of arithmetic, we may leave aside the controversies over mathematical concepts and focus on the meanings of the logical constants. In order for this controversy over the meanings of the logical constants to be deep, so as to represent a discussion with the program of militant intuitionism, each school should consider faulty those meanings of the logical connectives and quantifiers that the other group advocates. Consequently, the militant intuitionist must argue that it is impossible to ascribe to the logical constant those meanings that the classical mathematician (or logician) attempts to assign to them. The classical mathematician may make an apparent conciliatory move and apart from the classical constants introduce intuitionistic ones, marking them with different symbols. Moreover, as he may be interested in whether or not a proof or an inference is constructive, the new set of symbols may be of value for him, as they can be used to mark those constructive procedures that are agreeable to his intuitionistic rival. It is clear, however, that from the point of view of the militant intuitionist this move is hardly satisfactory. Our intuitionist will argue that his classical opponent has no right to use the classical logical connectives and quantifiers since their meanings are incoherent.

Thus, the question we are concerned with is, What are the *correct* meanings of the logical constants, as they figure in mathematical statements? The issue of correctness is relative to the subject matter discussed, or to the language of discourse. Now, the problem is how, or on what grounds, one may argue for a definite position concerning what meaning an expression may have. Obviously, to this end we need a viable concept of meaning applicable to sentences and their constituents. We need, moreover, an argument showing that meanings are not sacrosanct, so that the fact that the use of some expression in a linguistic community is well-entrenched and moreover regulated by some perceptible rules, does not prohibit criticism of its meaning. Finally, even if we had such an argument as well as an account of meaning which of necessity would be very general, how should we derive from a criticism directed against the meanings of the logical constants any conclusion about which putative rules of inference or logical laws are correct? The relation between meaning and forms of reasoning, or, to put the matter more methodologically, between an account of meaning and justification of forms of reasoning is in Dummett's project mediated by semantic theory, which has a twofold role. The first is the standard task, as acknowledged by logicians, of justifying the forms of reasoning. The other role is not so easily acceptable to the majority of logicians for whom semantic theory is merely a mathematical tool, useful for characterizing logics by means of some technical notions. Dummett requires namely that the semantic theory serve as a basis on which the related account of meaning is built. In other words, a given semantic theory should provide essential categories in terms of which the related account of meaning is to be given. Accordingly, what Dummett proposes is a rigid link between an account of meaning and its underlying semantic theory. Further, by relating meaning to knowledge, or understanding of language, he makes the account of meaning susceptible to criticism that draws, to some extent, on empirical data. For instance, one possible objection of this sort to an account of meaning can be that it makes meaning essentially incommunicable. In a similar vein, one may accuse an account of being in conflict with the empirically given fact of the acquisition of new meanings as we learn new expressions or become fluent in languages. The objections mentioned here form the crux of Dummett's reasoning for replacing, in mathematical discourse, classical logic by intuitionistic one, as they occur in his (PBIL) and (EI). There are also other possible criteria for the correctness of an account of meaning that do not appeal much to facts, but rather to methodological principles that prescribe what shape a satisfactory account of meaning should have. But, as we see it, the force of objections of the latter sort is rather limited.

Now, what emerges from the above picture of relationships among logic, a semantic theory, an account of meaning, and the understanding of language is a possibility of criticizing an inferential practice. One starts with logic, properly so called, and with its rules of inferences or/and axioms. In the second step, the logic is subjected to justification by a semantic theory. Then, the semantic theory that does this job is a candidate for being the basis of an account of meaning. Finally, given that the account of meaning is proposed, it may be criticized in respect to facts or methodological principles, as meaning is assumed to be on a par with understanding. If such a criticism is serious enough, it casts doubt on the semantic theory and indirectly on the validity of the accepted forms of reasoning. In principle nothing prohibits proceeding the other way round, that is from a satisfactory account of meaning *via* semantics to a position about which forms of reasoning are correct.

A successful execution of Dummett's project relies on establishing some rather recondite claims. Thus, to get a clear view of his argumentation, we need first to understand and second to evaluate these theses. We shall concentrate on the following issues:

1. What requirements, according to Dummett, should be fulfilled by a satisfactory semantic theory?
2. How does he understand the concept of meaning and what should the account of meaning explain?
3. In what sense should an account of meaning be based on a semantic theory?

Apart from these rather general issues stemming from the peculiarities of Dummett's program, we shall focus on queries more directly related to our project. These are:

1. Is it really so that an account of meaning based on a semantic theory that justifies classical logic is irreparably incorrect?
2. Is it so that the account of meaning that Dummett advocates leads, *via* the semantic theory on which it is based, to intuitionistic logic?

These are the questions we shall deal with below. We begin with an exegesis of Dummett's views on semantic theories, meaning, and the relation of the two, and then proceed to his central argument.

2. DUMMETT ON SEMANTIC THEORIES

2.1. *Three Tasks of Semantic Theories*

As Dummett sees it, the crucial notion of semantics is truth, although on the other hand the question of which concept of truth is correct falls outside its scope. (LBM, p. 21) On a pre-theoretical level we have a notion of valid argument, whose essential component is the idea that a valid argument leads from correct premises to a correct conclusion. The first move towards characterizing the validity is to focus on forms of argument rather than on particular arguments, whereas the second step is the momentous decision as to whether to characterize the validity in terms of truth, or otherwise.[45] We shall concentrate on the former characterization, aiming to disclose Dummett's requirements of what semantic theories should satisfy. To characterize which forms of argument are valid, a form of argument should be represented with the help of schematic letters. When we work with sentential logic, the letters are assumed to represent actual sentences. In predicate logic we need letters representing predicates (with a device marking the number of arguments they have), individual constants, and a means, usually letters for variables, of forming quantified sentences. Additionally, to allow for singular terms other than variables and singular constants, schematic letters representing functions used for forming complex terms are also introduced. A concentration on *forms* of arguments gives rise to a pre-theoretical concept of *interpretation by replacing schematic letters* (as Dummett calls it) and the characterization of the validity of the form of argument in terms of truth under interpretation. (LBM, p. 23) In the case of sentential logic, an interpretation of this sort is effected by simply replacing schematic letters by actual sentences; accordingly, a form of argument is valid if under any interpretation under which its premises come out true, the conclusion comes out true as well. In predicate logic, apart from stipulating which actual expressions of a suitable category may replace any given schematic letter, a range of quantifiers, that is, a domain, is also specified. Validity of forms of argument is characterized as in the case of sentential logic, that is in terms of truth under interpretation by replacement. Thus, in this pre-theoretical approach we appeal to some intuitive notion of truth. As Dummett observes, however, although sentences are assumed to be true or otherwise, we have no resources on this level to explain how the truth of a sentence depends upon the way it is composed. This task, "analysis of the way in which a sentence is determined as true or otherwise in accordance with its composition" (LBM, p. 24) is, as he points out, the principal task of any semantic theory. In his terminology, we say that the *semantic value* of an expression is the feature of the expression on which the truth of any sentence in which the expression occurs is dependent. Now, instead of replacing schematic letters with actual expressions and then assigning them semantic values, semantic theory directly ascribes to schematic letters those semantic values which it selects as appropriate for each category of schematic letters. Obviously, the choice of semantic values, the notion of truth, and the way the semantic value of a compound expression depends on semantic values of its constituents varies from semantic theory to semantic theory, although the last factor is to a great extent guided by the syntax the theory assumes. Working with Fregean syntax, as Dummett does, if we know for instance what semantic values of sentences are and that semantic values of singular terms are objects, it follows that the semantic value of an n-ary predicate must be a mapping from n-tuples of objects to semantic values assignable to sentences. (LBM, p. 27) This, however, is rather a knowledge of a general *form* of the semantic values that are assignable to predicates, since there still may be a controversy over what mappings are allowed, or what counts as an object. For example, as Dummett points out, contrary to the classical logicians who agree on arbitrary mappings, the intuitionist will opt in accord with his overall philosophical position for effective mappings only. A similar situation is faced in the case of the logical constants. Given that it is decided what semantic values are appropriate for singular terms and sentences, something is also known about what the semantic values assignable to the logical constants can be. That is, it is known that these are mappings of such and such a form.

Among various categories of semantic values, the semantic value of the sentence, or as Dummett calls it, *statement value*, has a significant role to play. It is the feature of the sentence on which its truth depends. It is instructive at this point to take an overview of Dummett's examples that attempt to show how the truth of the sentence under interpretation may depend on the semantic value of that sentence.[46] In the most familiar case of classical semantics, the statement value is simply identified with the truth or falsity of the sentence. Semantic theories that do not make this identification face the additional task of specifying how the truth of the sentence derives from its semantic value. For instance, Heyting's interpretation of the logical constants suggests that in intuitionism the statement value is identified with a principle that classifies constructions into those that prove the sentence, and those that do not. Then the notion of truth is arrived at by putting it that a sentence is true if a construction is known such that the principle classifies it as proving that sentence. Another example of introducing truth by means of existential quantification is that of Hintikka's theory, in terms of games, in which the statement value is identifiable with a set of successions of moves following an utterance of the sentence, the sentence being true if there is a winning strategy whose starting point is the announcement of this sentence. Another pattern of the relation between statement value and truth, not differing much from that of classical semantics, is provided by semantics for many-valued logics. Here the statement values are assumed to be elements of some set of cardinality higher than 2, whereas some elements (usually one) of this set are selected as 'designated'. The sentence accordingly is said to be true under an interpretation if the interpretation assigns to it a designated element. Still another link between truth and the statement value is assumed in those semantic theories that operate with relativised truth values. They assume a space equipped with a partial ordering, and require that at each point of the space all sentences, or at least all atomic sentences, are either true or false. Examples of this sort of semantic theories are: possible world semantics, with possible worlds as points and a relation of accessibility, Beth trees and Kripke trees with points interpreted as states of information ordered in accord with a relation of comprising, or semantics for a tense logic with a set of times, naturally ordered. Now, it is said that the sentence is true *simpliciter* if it is true at some selected point. This may be the actual world, the present state of information, or the present time. (LBM, pp. 34–35)

The way semantic values are introduced is to guarantee that the semantic value of a composite expression is determined, given the expression's composition, by the semantic values of its constituents. The task of a semantic theory splits accordingly into three parts: first, it should stipulate what the interpretation consists in, that is, what sort of semantic values are to be assigned to each category of schematic letters; secondly, it should show how the semantic value of a compound expression depends on the semantic values of its constituents; finally, it should state what relation holds between the semantic value of a sentence and that sentence's truth under the interpretation. (LBM, p. 35) We may call these tasks: the selection of semantic values, dependence of semantic values, and the relation between statement value and truth, respectively. To distinguish those semantic theories that fulfill these tasks from others that do not, let us call the former *genuine*.

There is still another requirement that Dummett puts forward as an ideal for a preferred semantic theory to satisfy, which stems from two postulates. First, in a dispute between adherents of different logics, each of the parties involved should be able to understand the meanings of the logical constants that the opponent advocates. The second stipulation requires that the dispute should be resolved, if possible, by assessing which of the two semantic theories that the protagonists propose is correct, where correctness is in turn judged by comparing the merits and demerits of the two accounts of meaning erected on these semantic theories. To come to terms with this ideal, we may imagine a dispute between a classical logician and an intuitionist who attempt to back their position as to the validity of some laws of sentential logic by an appeal to classical truth-tables. The classical logician observes that his opponent accepts each clause of the truth tables; for instance, he accepts the clause that $a \vee b$ is true only if a is true or b is true, though of course he means by this that either a proof of a is known, or a proof of b is known. Similarly, the intuitionist does not object that ¬a is true only if a is not true, though again, he construes this as requiring that a proof deriving a contradiction from the assumption that a holds is known. Thus, were the intuitionist asked by his opponent why, while accepting the classical truth tables, he rejects the principle of double negation elimination, or the excluded middle, he would reply that the way he understands negation and disjunction, as they figure on the right hand side of the clauses, does not permit him to assert some instances of these principles. Thus, the decision as to what laws are assumed to be valid in logic crucially depends on the laws of logic assumed to hold in the metalanguage. If the laws governing the logical constants in the metalanguage are classical, then truth tables validate classical logic and it can be shown that their lines exhaust all the possible cases; if the logical constants in the metalanguage are intuitionistic, the resulting logic will be intuitionistic. Such dependence of the laws of logic holding in the language on the laws assumed in the metalanguage in which the semantic theory is stated is expressed by saying that semantic theory is *sensitive* to the logic accepted in metalanguage. (LBM, p. 55) In the case of a sensitive semantic theory, the dispute over the validity of some laws of logic trivially shifts to a controversy over whether these laws hold in the metalanguage. We may put this in terms of meaning or understanding. The classical logician may be baffled why the adherent of some other logic rejects this or that law while being persuaded to accept classical truth tables for the logical connectives. Then, however, while attempting to find out why his opponent denies that a law of logic holds in the language, the classical logician inevitably discovers that he rejects this law to hold in the metalanguage, and of course this gives him no information whatsoever on his antagonist's reasons for denying the validity of the law in question. (LBM, p. 36)

Thus, there emerges the postulate that semantic theory should be as insensitive as possible to the logic assumed to hold in the metalanguage. (LBM, p. 55) Given a controversy between adherents of two logics, it would help to resolve the debate if all the laws holding in the metalanguage were non-controversial for both sides of the dispute. Thus, whether or not the disputed laws are assumed to hold in the metalanguage should have no bearing on whether or not these laws turn out to be valid in the language. The semantics of Beth trees (or Kripke trees) closely approaches this ideal. No matter whether the logic of the metalanguage is classical or intuitionistic, the same rules and laws of logic are validated, and moreover these are rules and laws of intuitionistic logic. So, if only the classical logician were persuaded to prefer Beth trees over classical semantics, the intuitionist would stand a great chance of winning the debate.

More realistically, both parties will propose different semantical theories. Given that party A defends the validity of some laws of logic that party B inveighs against, in the justification of these laws A should use in his semantic theory only those laws that B accepts. This ideal is not attained in model-theoretical semantics, since the justification of the laws that are controversial from the intuitionistic standpoint, say the principle of the excluded middle, draws heavily on the same, or others laws that are the subject of the controversy. On the other hand, if party A attempts to convince his opponent of the invalidity of a law of logic, then ideally he should do it within a semantic theory that permits this to be done by appealing only to those laws of logic that the other accepts. Obviously, in most cases B will not accept the semantic theory that his opponent puts forward. Nevertheless, this theory will help him see what rationale his opponent has for rejecting the disputed law. Thus, in the ideal case, both parties in a dispute over the validity of some laws of logic propose semantic theories that satisfy the above two conditions; the disagreement is then over what semantic theory is correct for a given language, for instance, the language in which mathematical reasonings are presented. As each of the proposed semantic theories should in turn provide a basis for an account of meaning for that language, they can be assessed as correct or not by identifying the meaning of expressions with understanding of them and testing whether the account accords with the facts of linguistic communication or, perhaps, with some general requirements imposed on the explanation of meaning. (LBM, p. 55)

2.2. *Programmatic Interpretation*

A semantic theory which satisfies the three requirements of the genuine semantics, that is, which selects semantic values, shows how they depend on each other and explains the relation between statement value and truth, should be distinguished from one that fails to do so. Clearly, without a statement of what sort of semantic values are to be associated with expressions of each category, we may have stipulations of how the truth or falsity of the compound sentence depends on the truth or falsity of the constituent sentences. Similarly, without making explicit what counts as an object, it may be stated under what condition atomic sentences are true: the notion of object is then purely programmatic. Such a programmatic interpretation, as Dummett calls it, does not show how the semantic value of a compound expression is determined by the semantic values of its constituents since it is not specific as to what semantic values are assignable to each category of expressions. It merely states how the truth of a sentence depends on the truth of the constituting sentences. (LBM, p. 32) A typical straightforward stipulation of this sort is, for instance, the following:

$a \vee b$ is true if and only if a is true or b is true.

In a similar vein a programmatic interpretation may straightforwardly stipulate the truth-condition of an atomic sentence whose principal operator is that of identity:

$a = b$ is true only if the terms a and b stand for the same object.

Straightforward stipulations are most familiar from truth tables, but they also occur in other semantic theories, like for instance those of Beth trees or Kripke trees. In these cases they are framed in terms of a relativised concept of truth, that is, *truth at a node*. Intuitively speaking, the sentence's being true at node t represents that sentence being verified at state t of information. Both semantic theories appeal to the straightforward stipulation to state what it means that the sentence with conjunction as the principal connective is true at a node:

$a \wedge b$ is true at node t iff a is true at node t and b is true at node t.

In the case of other connectives, however, the stipulations are not straightforward. For instance, we have:

$a \Rightarrow b$ is true at node t iff for every $s \geq t$, if a is true at node s, then b is true at node s.

Now, the programmatic interpretation, that is, the interpretation appealing *only* to straightforward stipulations, is maximally sensitive to the logic assumed in metalanguage: whatever laws of logic are obeyed by metalinguistic *and, or, if, no, every, some*, they turn out valid for object-language connectives of $\wedge, \vee, \Rightarrow, \neg, \forall, \exists$, and similarly each law of logic valid in the object language has an exact counterpart among the laws holding in the metalanguage.

One may, however, doubt whether there really is such a gap between genuine and programmatic interpretations as Dummett makes out. First, it may be believed that the purported gap disappears once it is specified, against the background of a programmatic interpretation, what the semantic values of each category of expressions are. Secondly, given a language with classical logic, one may claim that the difference between the programmatic interpretation and the semantic theory shrinks to triviality. To address the first issue, let us consider a programmatic interpretation for intuitionistic logic (Dummett dubs it *internal interpretation*). (LBM, p. 56) It consists, first, in the specification of an inhabited domain, this being an intuitionistic *species* with a known constructive proof that the species comprises at least one element. Secondly, the interpretation assigns to individual constants—elements of that species, to each monadic predicate—a subspecies of the domain, in general, to each n-ary predicate—a species of n-tuples formed out of elements of the domain. Thirdly, truth-conditions of atomic sentences are stated in a straightforward way. Thus, for a monadic predicate F, a clause states that the sentence $F(a)$ is true under the interpretation only if an object associated with a by this interpretation belongs to the sub-species of the domain that is assigned to predicate F. Fourthly and most importantly, as to compound sentences, their truth-conditions are stated in a straightforward manner, in terms of the truth of their constituents. The result of the last stipulations is that, if the logical connectives occurring in them are understood intuitionistically, the internal interpretation identifies as valid only the intuitionistic laws of logic, whereas, on the other hand, if they are ascribed classical meanings, the resulting logic is classical. Thus, in order for this to be an interpretation of *intuitionistic* logic, we have to assume the intuitionistic connectives in the language in which the stipulations are stated. (LBM, p. 56) Now, the question is what semantic values should be assigned to sentences? Although the internal interpretation assigns to the monadic predicate F a sub-species of the domain-species, the object assignable to the sentence $F(a)$ remains unspecific, at least as long as the notion of species is unclarified. Thus, it is tempting to say that as soon as a decision is made about what sort of things the statement values are, the internal interpretation will turn into a genuine semantic theory. After all, the internal interpretation states what the domain is, what should be assigned to individual constants and predicates, and how the truth of compound sentences depends on the truth or otherwise of the constituent sentences. Thus, as a result of the decision, we will have the stipulation of what the statement values are plus conditions stating how the truth of the compound sentence depends on the truth of the constituting sentences. How then will these two tally with each other, that is, will we automatically obtain an explication of how truth relates to the statement value and how the semantic values of compound expressions depend on those of their constituents? One may hope that, granted that statement values are selected for *atomic* sentences, the rest can be explained in terms of truth. This, however, does not hold as Dummett shows. (LBM, p. 58) Suppose that against the background of the internal interpretation for intuitionistic logic, one assigns those statement values to *atomic* sentences that Beth tree semantics prescribes. That is, each atomic sentence is said to be either true at a node or not true at a node. Moreover, it is postulated that truth *simpliciter* be identified with truth at the vertex, that is, a node representing the present state of information. Now, guided by the straightforward stipulations, we attempt to explain the condition for, say, the negated sentence to be true under an interpretation with respect to Beth tree. This amounts to saying that

$\neg a$ is true *simpliciter* iff a is not true *simpliciter*.

This, however, does not hold since the implication from right to left is invalid, because in order for $\neg a$ be true at the vertex, a must be not true at all nodes bellow. Clearly, the equivalence does not hold because in the semantics of Beth trees, negation does not commute with truth at a node. One may tend to ignore this point by arguing that it should always be possible to coin such a concept of truth that would commute with negation, and perhaps, distribute over other logical operators. However, as Dummett insists, there is no guarantee that that concept of truth, as applicable to *atomic* sentences, could coincide with that introduced in terms of the factors conceived of as semantic values of expressions that constitute the atomic sentence. (LBM, p. 58) Thus, the mere selection of what statement values are does not establish a semantic theory, as it does not show how the semantic value of a compound sentence is to be determined by semantic values of sentences that constitute it and leaves unspecified the relation between the semantic value of the sentence and the truth of that sentence.

Now, we need to look into our second problem, namely the relation between a programmatic interpretation for a language with classical logic and a semantic theory for this language. The question is whether one may dispense in this case with the latter, hoping that it will be possible to extract basic semantic principles from the programmatic interpretation. The salient feature of classical semantics being that the statement value is simply identified with the truth or falsity of the sentence, straightforward stipulations for the logical constants seem to state how the semantic value of a compound sentence depends on the semantic values of the sentences that constitute it. It thus appears that, without taking a stand on what statement values are, one may *derive*, granted classical logic is assumed in the language in which the stipulations are stated, the classical truth table for sentential logic, together with the principle that their lines represent all possible cases. (LBM, p. 62) Some other principles characterizing the classical logic of the language considered can also be derived from the straightforward stipulations and laws of classical logic that are assumed to hold in the metalanguage. It appears also that it does not matter that we do not know what should be assigned to sentence-letters; we may call that thing 'proposition' and all that we need to know about it is that it may be said to be true or not true. Similarly, given only the programmatic interpretation, we may content ourselves with the knowledge that a property should be associated with a monadic predicate letter. And of a property we should only know that each object of the domain either has it or has not. Now, the objection goes, if one starts with a programmatic interpretation that does not make explicit the above 'facts' concerning propositions and properties, one may derive them by assuming classical logic to hold in the metalanguage in more or less the same way as one derives classical truth tables for sentential logic with the principle that their lines exhaust all possible cases. That is, by assuming that 'A is false' means the same as 'A is not true', one derives the principle of bivalence; then by proving that sentences having the same truth-value are replaceable within any complex sentence without changing the truth-value of the whole sentence, one shows in effect that the semantic value of a sentence may be identified with its truth-value. Similarly, one recovers the principle that the semantic value of a unary predicate is a set by showing that predicates of the same extension are intersubstitutable. If that is really so, a programmatic interpretation is all one needs to account for a language with classical logic, an explicit statement of the principles of classical semantic being an unnecessary ornament. This would mean that for a language with classical logic, the gap between programmatic and semantic interpretations shrinks to that between what is implicit and what is fully extracted. (LBM, p. 73)

To dispel this objection Dummett's tactics consist in, first, demonstrating a language with the classical logic for which the straightforward stipulations do not achieve anything comparable to a semantic theory and, second, arguing that the derivation of principles of classical semantics from the classical logic assumed in metalanguage is spurious, as it leads to weaker results than those characterizing classical semantics. We will concentrate only upon the first claim. To establish it, Dummett invites his readers to consider a language containing vague predicates whose functioning is nevertheless understood as governed by classical logic. The predicate is vague if it definitely applies to some objects and definitely does not apply to some other objects, there being some objects to which it neither definitely applies nor does not definitely apply. Accordingly, an ascription of a vague predicate, say 'bald', to an individual may be correct, to another—incorrect, there being some borderline cases for which it is neither definitely correct nor definitely incorrect. It thus appears that sentences formed by applying a vague predicate to names of objects that lie on the borderline may not respect bivalence. However, given an object a whose color lies somewhere on the borderline between pink (P) and red (R), there is an inclination to say that the alternative $P(a) \lor R(a)$ is true which, moreover, given that we conceive of these two predicates as excluding each other, justifies the truth of the instance of the excluded middle, that is $P(a) \lor \neg P(a)$. These intuitions, pointing to the possibility of a language in which bivalence fails, while sentences of that language nevertheless obey classical logic, are accommodated in the semantics of supervaluations, the basic idea being the analysis of various ways vague expressions can be made precise. First, one considers a space consisting of a non-empty set of points, with relation $\leq$ of partial ordering, where the point represents a step in making vague predicates more precise (a *precisification* of predicates) and the ordering corresponds to the intuitive relation of increasing precision. The space thus represents various ways in which language can be made more precise, and accordingly, the ways indeterminate sentences can be made true or otherwise. Of the space it is required that it have a base-point representing the 'received state of vagueness', and that for any point u there be some point $t \geq u$ representing a complete precisification of expressions, that is, one allowing for no indeterminate sentences. It is further required that any indeterminate sentence be resolvable in both ways, that is, that there be both precisifications under which the resultant sentences come out true and precisifications under which they come out false. Moreover, it is required that:

1. a sentence is true (false) at a point of complete precisification if and only if it is classically true (false);
2. subsequent precisifications should preserve truth-values.

The two conditions are sometimes called *fidelity* and *stability* respectively. (Fine 1975, p. 272) The process of making predicates more precise should be rather subtle if we intend to uphold disjunctions like 'pink(a) or red(a)' where *a* lies on a borderline between pink and red, so that neither pink(a) nor red(a) is definitely true. It cannot happen that at a point of complete precisification 'pink' and 'red' are made precise in such predicates P and R, respectively, that leave things that are on the borderline neither P nor R. Now, to introduce a concept of truth to this framework, it is stipulated that the sentence is true if any sentence obtained from it by replacing its vague predicates by predicates assigned to them at any point of complete precisification comes out true. As Dummett stresses, the way truth is introduced guarantees that sentences with vague predicates obey the same laws of logic as that fragment of the language in which no vague predicates occur. Thus, if for some reason classical logic is preferred for the fragment of that language that contains only non-vague predicates, the whole language has classical logic, although bivalence clearly fails. What semantic values, then, are to be assigned to sentences under this semantic theory? An objective feature that attaches to a sentence in that language is plainly not its being simply true or not, but rather a distribution of truth-values of resulting sentences, which may be represented by an ordered pair $\langle U, V \rangle$, U being the set of all points of complete precisification and V— set of points of complete precisification at which the resulting sentences come out true.

Now, the programmatic interpretation for that language does not differ from the programmatic interpretation for a language without vagueness. Thus, as Dummett observes, appealing to classical logic one may derive that the logical operators satisfy the truth tables. (LBM, p. 73) Further, one may obtain that predicates are intersubstitutable provided they have the same extension. The programmatic interpretation will not show, however, what the extension of the vague predicate is, the extension being somehow framed in terms of points of precisification. Much more importantly, the concept of truth, as it is used in the semantic theory for that language, does not tally with the notion of truth that occurs in straightforward stipulations. Given that (absolute) truth of a sentence is identified with all resultant sentences coming out true at all points of complete precisification, absolute truth does not commute with negation since it does not follow from the fact that A is not true that $\neg A$ is true. More generally, absolute truth does not distribute over logical operators, and accordingly, it is not the concept of truth that occurs in the straightforward stipulations of the programmatic interpretation. (LBM, p. 74) Thus, in the case discussed, the programmatic interpretation shows neither what semantic values are associated with expressions of each category, nor how semantic values of compound expressions are determined by those of their constituents, nor what the relation is between the semantic value of a sentence and truth. This shows that even in the case of a language with classical logic, the programmatic interpretation fails to satisfy the requirements on semantic theory.

So far we have been discussing Dummett's views on semantic theories and in particular his arguments for distinguishing between genuine semantic theories and merely programmatic ones, with the former aiming to capture the functioning of language and the latter failing to do so. The genuine semantic theory, to repeat, handles three tasks: it assigns semantic values to each category of expression, explains how the semantic values of the compound expression depend on the semantic values of its constituents, and finally states what the relation between the statement value and truth is. Below we shall need to consider why a genuine semantic theory for a language, with genuine interpretation, is significant for framing an account of meaning for that language. Here, however, we need to postpone these investigations in order to clear up some possible confusion that might have resulted from our mention of Beth tree semantics whenever the controversy between intuitionistic and classical logics has been discussed.

2.3. *Skeletal Semantics*

An impression might have arisen that Beth trees provide the semantic theory for intuitionistic logic that Dummett seeks, the only remaining question being the choice between this theory and classical semantics. Dummett holds, nevertheless, that it is not so; the semantics of Beth trees still falls short as a foundation of an account of meaning. It is instructive to see his reasons, as they point to additional requirements the genuine semantic theory should satisfy.

In the case of classical semantics, there seem to be mutual relations between the intuitive meaning of a formula and its interpretation. For instance, given that we know the meaning of a formula containing a binary predicate, we may construct its interpretation by stating the domain and laying down for which ordered pairs of elements of that domain the formula comes out true. But we also gain understanding of a formula by being told what the domain is and for which ordered pairs of elements of the domain the formula comes out true. More generally, finding a possible extension of an *n*-ary predicate seems to suffice for grasping its meaning. Thus, as Dummett remarks, in the case of classical semantics it appears absurd to ask, given that the extension of a predicate is known, what that predicate actually is. (LBM, p. 151) Additionally, the idea that the semantic representation fixes the intuitive meaning tallies well with a more general principle that Dummett attributes to Frege, and which requires that

(...) a stipulation of what the semantic value for an expression is to be confers a particular sense on that expression. (LBM, p. 148)

Dummett explains this principle by invoking Wittgenstein's distinction between saying and showing. The account of meaning contains clauses that say (or state) what semantic values are assigned to expressions; on the other hand, the very way these stipulations are presented shows (or displays) what sense is attached to the expressions considered. This does not mean, however, that the explanation of senses of expressions must always take the form of a stipulation of what semantic values the expressions have; rather, whenever such stipulations are a part of the explanation of meaning, they confer particular senses on the expressions considered. Now, as Dummett observes, the principle is violated in the semantics of Beth trees. His argument reflects on de Swart's demonstration of a Beth tree such that at its vertex the formula $\neg \forall_x (F(x) \vee \neg F(x))$ comes out true, the domain being the set of natural numbers.

To come to terms with the semantics of Beth trees, let us imagine the mental activity of an idealized mathematician as being divided into stages representing that person's states of knowledge. It is assumed that knowledge increases without any loss as one proceeds to the next stages. At each stage the subject may establish some truths and have some options of how further to extend his knowledge. This is represented by a fork, with more than one stage following immediately after a given one. Consequently, the subject's possible growth of knowledge is viewed as an infinite tree, whose nodes represent states of knowledge and each path through the nodes—a possible development of his knowledge. Finally, the vertex of the tree portrays the present state of knowledge of the idealized mathematician.

How should we interpret the subject's knowing at node α that sentence ψ is true? It appears that the demand that sentence ψ be actually verified at node α is too restrictive, the more recommendable option being that it suffices if the subject will verify ψ, no matter in which way he further develops his knowledge. To express this more precisely, we need the concept of *bar*: a bar for node α is a set B of nodes following α such that each path through α intersects B. Now, we may say that the subject knows the truth of ψ at node α if there is bar B on α such that at each $\beta \in B$, the subject verifies ψ. This heuresis underlies the following formal definition:

A Beth tree is a quintuple $T = \{M, \leq, \chi, D, \vdash\}$ where:

M is a set of nodes partially ordered by $\leq$, χ is an assignment function that associates with each atomic sentence a set of nodes (intuitively, those at which the sentence is verified), D is a set of objects (domain) and $\vdash$ is a relation between elements of M and sentences—it represents the sentence being true at a node.

As to the sentence being verified at a node, it is required that for any nodes α, β and any atomic φ

1. $\alpha \in \chi(\varphi)$ or $\alpha \notin \chi(\varphi)$, that is, φ is verified at node α or is not verified at α;
2. if $\alpha \leq \beta$ and $\alpha \in \chi(\varphi)$, then $\beta \in \chi(\varphi)$, that is, if φ is verified at node α, then it is verified at any subsequent node.

Truth at a node under assignment χ is then inductively defined as follows:

1. $\alpha \vdash \varphi$, for φ atomic, if there is a bar B for α such that $\forall_{\beta \in B} \beta \in \chi(\varphi)$;
2. $\alpha \vdash \psi \wedge \varphi$ if $\alpha \vdash \psi$ and $\alpha \vdash \varphi$;

3. $\alpha \vdash \psi \lor \varphi$ if there is a bar B for α such that $\forall_{\beta \in B} \beta \vdash \psi$ or $\beta \vdash \varphi$;
4. $\alpha \vdash \psi \Rightarrow \varphi$ if $\forall_{\beta \leq \alpha}$ if $\beta \vdash \psi$, then $\beta \vdash \varphi$;
5. $\alpha \vdash \forall_x \varphi(x)$ iff $\alpha \vdash \varphi(a)$ for every a from D;
6. $\alpha \vdash \exists_x \varphi(x)$ if there is a bar B for α such that $\forall_{\beta \in B} \exists_{b \in D} \beta \vdash \varphi(b)$;
7. $\alpha \vdash \neg \varphi$ if $\forall_{\beta \geq \alpha}$ it is not so that $\beta \vdash \varphi$.

Finally we say that the closed formula φ holds (is true *simpliciter*) in Beth tree $T = \{M, \leq, \chi, D, \vdash\}$ if $\alpha \vdash \varphi$ for all $\alpha \in M$. This is equivalent to φ being true at the vertex.[47]

The tree de Swart (1974) demonstrated is binary, with the set of natural numbers as the domain. Its nodes may thus be represented as finite sequences of 0's and 1's, with 0 and 1 imagined graphically as 'moves' to left and right, respectively. It is then stipulated that the sentence $F(m)$ is verified at node α iff α contains at least $(m+1)$ 0's, which means that to reach this node one needs to take at least $(m+1)$ moves to the left, not necessarily in succession. This has the consequence that the sentence $F(n) \lor \neg F(n)$ is not true at any node of level n, that is, one containing exactly n 0's and 1's. For, in order for this sentence to be true at node α of level n, α should be barred by a species B such that for any $\beta \in B$, $F(n)$ is true at β or $\neg F(n)$ is true at β. Consider now a node γ lying on the utmost right path from α. $F(n)$ is not true at γ since there is no bar for γ such that at every node from it $F(n)$ is verified. On the other hand, for $\neg F(n)$ to be true at γ, at every node following γ $F(n)$ must not be true. However, $F(n)$ is true at nodes that can be arrived at from γ by taking $n+1$ moves to the left, not necessarily in succession. Thus, $F(n) \lor \neg F(n)$ is not true at α. Finally, since at no node of level n is the sentence $F(n) \lor \neg F(n)$ true, it follows that at no node of this level is the sentence $\forall_x(F(x) \lor \neg F(x))$ true, and as this obtains for every level, $\neg \forall_x(F(x) \lor \neg F(x))$ is true at the vertex.

Now, as no arithmetical predicate is known for which $\neg \forall x(F(x) \lor \neg F(x))$ is true, the description of the above tree falls short of specifying an actual predicate. The tree may be seen as stating some conditions that the desired predicate should satisfy; yet these conditions do not permit picking out an actual predicate. Even more, it is conceivable that no predicate satisfies them. The conditions attempt to state what F is by means of stipulating which complex sentences involving F are assertible and which are not. They require that (1) the verification of $F(m)$ proceeds after the verification of $F(n)$, for $n < m$, that (2) no statement of the form $\neg F(n)$ is ever true and that (3) $\forall_x F(x)$ is never true. (EI, p. 415) As the conditions are rather simple, one may hope to construct the desired predicate by building formulae of appropriate complexity that satisfy them. Yet, as Dummett shows, this makes us move along a circle (*ibid.*) Thus, although the Beth tree imposes restrictions on the sense of the predicate, it does not confer on it a definite sense. What is then needed to turn Beth trees into a semantic theory that fulfills the principle that stating the semantic value of an expression displays its sense? We definitely need more than an abstract structure of states of knowledge with a stipulation of what atomic sentences are verified at each. What is required is an explanation of what counts as verification of any given atomic sentence, which should permit the identification of the actual states of knowledge with the abstract nodes of the Beth tree. Given that we know what the information available at a node is that makes $F(n)$ verified there, we will automatically know what the predicate is for which $\neg\forall_x(F(x) \vee \neg F(x))$ is true. The fact that such a supplementation of Beth trees is required for allowing it to confer definite senses on the expression interpreted is the reason why Dummett considers it a sort of skeleton semantics. Obviously, the failure to live up to that standard of genuine semantic theory has little to do with the purposes Beth trees serve in logic, as the abstract structure of nodes suffices for a proof of completeness or soundness of intuitionistic logic.

Until now we have been dealing with an exposition of Dummett's views concerning semantic theories. We are now ready to investigate the relations between a semantic theory and an account of meaning, and specifically, the notion that the semantic theory is the basis for the meaning theory.

3. DUMMETT ON MEANING AND ITS BASIS

3.1. *Meaning, Knowledge and Understanding*

The point of departure of Dummett's characterization of what meaning is accords with common sense associations that relate meaning to knowledge and understanding. The meaning of an expression is that part of knowledge that constitutes an understanding of the expression; it is the knowledge in virtue of which it is said of a speaker that he understands the expression. (WTM, p. 69) In still other words, it is that piece of knowledge that must be possessed by someone who counts as a competent language user. (LBM, p. 83) The nexus between meaning, knowledge, and understanding appears rather intuitive—we commonly refuse to allow that someone knows the meaning of an expression on the grounds of his observed ignorance of some significant facts relevant to mastering the expression. For instance, if you observe a foreigner who, when addressed in English with a certain ichtyological problem, shows hesitancy as to whether fish live in water, you will most probably judge from this that the person does not know the meaning of the word 'fish'. On the other hand, however, it is plain that not all the knowledge pertaining to the expression goes into its meaning. The fact that only a few people know the average number of stripes on the legs of okapi does not indicate that the great majority are ignorant about the meaning of the word 'okapi'. Examples of this sort illustrate the need for a distinction between knowledge of language and extra-linguistic knowledge, with a further problem being who is a bearer of the knowledge of language. Another task that emerges from the above formulations is explaining the nature of the knowledge of language. In this chapter we will touch upon both these issues, and then proceed to Dummett's distinctions between ingredients of meaning because one of those, known in Fregean tradition as *sense*, plays the crucial role in Dummett's project. Before we proceed let us, however, straighten out our terminology and clear up a point of methodology. So far we have been using the neutral expression 'account of meaning' for what Dummett calls 'a meaning theory', that is a systematic representation of the meanings of all the words and expressions of a given language. (LBM, p. 22) This theory should be contrasted with what is dubbed 'the theory of meaning', which coincides with such branches of philosophy as theory of knowledge or philosophy of language. As to the relation between the two, the latter investigates what general principles or constraints should be observed in building a theory of the former type. The meta-theoretical level at which these investigations are located discloses a troublesome aspect of Dummett's approach. It is, as he hopes, the theory of meaning that is expected to clarify, or even solve the age-old disputes about what there really is, what the nature of truth is, what the sense of expressions consists in and so on. Yet the answers to, or (more modestly) the clarification of, these fundamental problems is to be achieved by reflecting on a possible form that a meaning theory may take. It is highly debatable, however, whether we have anything that deserves the name of meaning theory at the moment and it is not excluded that the nature of the matter, that is of knowledge of language, is such that it will never be adequately represented. It thus looks as if reflections on the principles of building some non-existent theories have dictated solutions to problems in metaphysics, philosophy of mathematics and the like. But if this is so, the persuasive power of conclusions reached on the grounds of such investigations will be meager. Even if our studies of the principles of constructing meaning-theories showed beyond a doubt that bivalence cannot hold for yet-undecided statements and moreover that the intuitionist is right in rejecting some laws of classical logic, it is still hard to believe that *this* will convert classical mathematicians to intuitionism. Fortunately, however, apart from the later incarnation of Dummett who approaches the matter in such a methodologically oriented manner, we have the earlier Dummett daring to argue more directly for intuitionism and other anti-realistic positions. We will thus concentrate on the earlier arguments, reflecting on the later only if they shed additional light on or lend support to the former.

Let us return to the issue of relating meaning to knowledge. If we concentrate attention on a given person's use of language, the envisioned relation is apparently contradicted by our readiness to ascertain that someone knows the meaning of a given expression, although that person does not know the criteria of application of the expression, or in general, even though his knowledge pertaining to the word in question falls short of an ability to explain, or to decide what it applies to. Such cases may be divided into two groups, according to whether there is a human authority on application of the expression or not. More perceptible is perhaps the first category as exemplified by Putnam's 'gold' example. This word is customarily used by variety of speakers in everyday situations and it is readily ascertained that a speaker grasps its meaning if he can tell, say, a gold ring from rings made from some other metals or is able to explain that gold is a precious yellowish metal. Yet only a few people, typically jewelers and metallurgists, know in a strong sense what gold is, as they are familiar with its chemical structure and tests permitting them to distinguish gold from fake gold. 'Gold' shares this feature with many expressions of everyday language that function as well in scientific discourse, like 'temperature', 'electricity', or 'combustion': they all belong to the first category. The category also contains less perspicuous cases, as shown by Dummett's example with the word 'colonel'. Evidently, this word may be used by a person who has no idea of the ordering of ranks in the military forces. There is an inclination to agree that in certain circumstances such a person understood the remark 'A colonel stepped out of the limousine', or that he repeated it with understanding, despite the fact that he could come up with only a lame explanation of what that word applies to. The example is less convincing since we are less inclined to say in such cases that the person knows the meaning. Nevertheless, what unites the 'colonel' example with cases like 'gold' or 'temperature' is the existence of acknowledged authorities on the application of the word, in the sense that every competent speaker's use of these words is held hostage to the authorities' application of them. The layman assents to a jeweler's verdict that a ring is not gold, though it looks like gold, or to a statement by someone well-informed on military ranks that a soldier is indeed a colonel. As we turn to the second category, on the contrary, there are no authorities on the use of expressions belonging to it, or at least they are not embodied in any concrete human beings. The most typical examples are those of geographical names or the names of historical or public figures. We will probably credit someone with a grasp of the meaning of 'Prague', even if he knows of Prague only that it is a big historical city somewhere in the Czech Republic, as he appears to understand someone's saying that she spent Christmas in Prague. Yet a question of what it is, that unique and complete knowledge that constitutes someone's full grasp of the meaning of 'Prague', appears to be ill-posed. Moreover, one can draw only an arbitrary distinction between which of the indefinitely many possible understandings of such a name is better, provided they allow the speaker to locate the denotation. A similar predicament is faced in an attempt to explain what minimal piece of knowledge allows one to be credited with grasping the meaning of a geographical or historical name. Consideration of this sort drives Dummett to emphasize the social character of language. One aspect of this is the existence of people acknowledged as authorities on applications of some words, such that every language-user aware of the language he speaks holds his use of expressions of a given kind to be hostage to what the authorities judge correct or not. Another aspect derives from the existence of custom, social practices, and institutions, and this is especially important for the use of words belonging to the second category. As practical abilities relevant for the functioning of these words, Dummett lists abilities to get to the named places, or telling in what place the person is, and that rests on a practice of drawing maps, abilities to read them, the existence of various means of transport with systems of signs, landmarks or road signs. Moreover, in the case of some names, typically of places of historical significance, the ability to locate the place on a map, or recognize it when seen, still may not count as a full grasp of its meaning; apprehension of its historical or cultural significance is also required. All this indicates that the use of such expressions is further influenced by relatively new institutions that collect information relevant to such places and pass it on to society, such as geographical dictionaries, travel agencies, tourist guides and the like. (LBM, p. 86)

Now, the acknowledgment of the social character of language leads to a loosening of the initially tight link between meaning and understanding. Instead of simply locating meaning in the heads of speakers, we need to view expressions of language as having meaning in virtue of belonging to a public language, spoken by a community that consists of both laymen and experts on applications of some words, a community in which various social practices and institutions exist. As to individual speakers, they are said to grasp or not grasp the meanings of expressions, and moreover, their grasp of meaning is gradable, the more complete the more the speaker knows about the kinds of context in which their use is appropriate. Nevertheless, in spite of the loosening of the relation between meaning and understanding, a methodological point about the two remains in force: the philosophical questions concerning meaning are best interpreted as questions concerning understanding; a query about the meaning of a given expression is best handled as pertaining to what it is to know its meaning. (WTM, p. 69) This formulation leaves untouched the task of characterizing what the knowledge of the meaning of the expression, or more generally, the knowledge of language is. In most of his writings, with the notable exception of LBM, Dummett seems to equate this with a practical ability, possibly of a vast complexity. In the book mentioned, this identification is replaced by the requirement that the knowledge of language, or of a given expression, should issue in various practical abilities. He points moreover to phenomena that seem to conflict with identifying knowledge of language with practical abilities, the crucial observation being that such identification makes it hard to explain that we often *know* that we understand an utterance. (LBM, p. 93) Here our readiness to answer, without any hesitation, various questions of the form: 'do you understand what was said?' is viewed as testifying to our knowing that we understand an utterance, though it obviously may happen that we are mistaken in thinking that we do so. Now, it is possible that someone may exercise a merely practical ability while remaining unaware of possessing it. We may conceive of a parent who, after being asked by a child, 'Can you draw me a horse?' replies: 'I don't know, I've never tried'. Obviously, for other cases of practical abilities, this kind of an answer will not be given, apart from responses intended as a joke like 'I don't know if I can swim; I've never tried.' It is a joke since we know whether we can swim, ride a bike, operate an electric drill as we are aware that people are not born with these abilities and remember having learned the respective skill, or not. This suggests that one may know what these skills are without knowing how to exercise them. It is intelligible that a person knows what it is to draw a horse, though he does not know how to do it. Thus, no matter whether the practical ability is such that one may or may not know that he can exercise it, there appears to be a distinction between knowing what a given practical ability is and knowing how to exercise it. Given that one knows what a particular ability is, it looks as if this person confronted only the task of achieving a certain technique that would enable him to exercise it.

Contrary to that, as Dummett argues, if you do not know how to speak a language, then neither do you know what it is to speak that language. To take his own illustration, it makes sense to say of a person that he knows what swimming is, although he cannot swim—that person can tell whether or not someone else is swimming and can even try to swim. (LBM, p. 94) On the other hand, it does not make sense to ascribe to someone who cannot speak Spanish the knowledge of what it is to speak Spanish, as that person can easily be taken in by anybody pretending to speak that language. Similarly, this person cannot even try to speak it, as he has no idea what to do to this end. It is not that a person knows in advance what it is to speak the language, and then gathers the information and learns the skills that will allow him to execute the ability. There are obviously some aspects of knowledge of language that fall under this pattern, for instance pronunciation, but they are rather the exception than the rule. In general, if one does not know a language, he does not know what it is to speak that language. Thus, in the case of knowledge of language, as Dummett concludes, the distinction between knowledge of what the ability is and how it can be exercised disappears or at least fades away. (*ibid.*)

Nevertheless, the asymmetry noted between knowledge of a language and knowledge of a practical ability appears thin, which may suggest that perhaps knowledge of language is after all a practical ability. Let us consider again if a person who does not speak Spanish can be taken in by those who pretend to do so. Clearly, that may easily happen, yet the same sort of joke may be played on someone who is illiterate, but is said to know what reading is: you may take him in by pretending to read. And the objection that the butt of the last joke would in principle be able to figure out whether or not he was being taken in, whereas someone taken in by pranksters pretending to speak Spanish would not, does not appear cogent. If you allow for a wider sense of 'being able' as it occurs in the phrases above (and you need to do this for the first case to hold), then clearly the latter person can find out whether or not others speak Spanish. If this person is credited with knowing what it is to speak Spanish, he must grasp the meaning of 'Spanish language', and this in turn presupposes his awareness of natives of what countries speak Spanish, which may issue in his ability to locate Spain or some Spaniards. Accordingly, when in doubt as to whether Spanish was spoken on a given occasion, the person knows how to find an authority on this question. Similarly, there is no unbridgeable gap between, on the one hand, the intelligibility of the supposition that a non-swimmer attempts to swim, and the meaninglessness of the supposition that someone who does not know Spanish tries to speak that language, where both persons have only knowledge 'what', and no knowledge 'how'. To cut a long story short, if it is a single try, then in both cases it is ridiculous and will inevitably be deemed a flop. But, if it is a sufficiently long series of tries, in water or in the environment of the Spanish language, then in both cases it can lead to the desired result. That is, if suitably placed, (and for finding this environment the knowledge 'what' is relevant), a person can turn his ever-so minimal knowledge of what speaking a given language is, into knowledge of that language. Thus, what Dummett's argument shows is not an unsurpassable gap between knowledge of language and knowing a practical ability, but rather a spectrum of practical abilities, with the knowledge of language at one of its extremes.

Dummett draws, however, a stronger conclusion, to the effect that knowledge of language is *not only* the practical ability, as it also involves theoretical knowledge. (LBM, p. 94) Theoretical knowledge consists in a capacity to formulate in a systematic way certain propositions concerning a subject matter, display logical relations between these propositions, derive their consequences and answer questions concerning them. (*ibid.*) Viewed in this way, however, only a few speakers have theoretical knowledge of a language and only on rather special aspects of it. Most speakers can state the meanings of some expressions in other words, but if they do not know the rudiments of grammar they can hardly formulate the rules they follow in using tenses, for instance. Hence, the claim that the knowledge of a language is *explicit* should be relaxed to a demand that the speaker be able correctly to recognize whether an explanation of a word is correct, or whether a grammatical rule correctly captures his linguistic practices. With this shift, from the ability to state to the ability to recognize, we turn to characterizing knowledge of language as *implicit* theoretical knowledge. Could the knowledge of the language then be equated with *implicit* knowledge of a meaning theory for that language? As Dummett observes, the concept of *implicit* knowledge poses the question of how much prompting and assistance is needed to make someone recognize that a given formulation correctly captures his use of language. But even if no limit is imposed, this does not make implicit knowledge elastic enough to ascribe to the speaker the implicit knowledge of a meaning theory. As Dummett sees it, if presented with that theory, the speaker would probably not understand it, and even if he were, he would start disputing its various claims, with the chances that these controversies would ever be resolved being slight indeed. (*ibid.*, p. 96)

What emerges from Dummett's discussion is a spectrum of modes of knowing the language whose endpoints are, on the one extreme, practical ability, and on the other, explicit theoretical knowledge. We learn a substantial part of the vernacular language by being given explanations, stated by means of words we already know, of unknown parts of the vocabulary. This part of knowledge of language falls under the category of theoretical knowledge (explicit or implicit). For obvious reasons, however, we cannot think of our entire linguistic knowledge as being gained in this way. Thus, its basis must be a practical ability. Now, this rough distinction has a bearing on explanations of meanings, or on the form of a meaning theory. Although for a number of expressions it suffices that the theory restate their meaning in different words, for other expressions it must be explained how the grasp of their meaning is manifested, where one may think of these two categories of expressions as corresponding, respectively, to the part of language the command of which constitutes theoretical knowledge and the other part, the command of which constitutes a practical ability.

Thus, starting with some intuitions concerning the notion of meaning, we have arrived with Dummett at a requirement for the form of a meaning theory. We were concerned with a link between meaning, knowledge, and understanding. Dummett's acknowledgment of social character of meaning relaxes this nexus, with the expression having meaning by virtue of belonging to the social language, meaning being no longer simply identified with a particular person's piece of knowledge. Despite the loosening of this relation, a methodological point remains in force that recommends translating queries about meaning into appropriate questions concerning knowledge or understanding. Finally, we have discussed Dummett's argument against identifying knowledge of language with a complex of practical abilities, coming to the conclusion that it leaves much to be desired.

3.2. *Sense, Force and Holism*

So far we have operated with an undifferentiated concept of meaning. But clearly, in order to use an expression, say a sentence, one needs to know various things, and that points to a task of distinguishing components of meaning. The first distinction of this sort that comes to mind is the one between the content of a given utterance and what goes into indicating the conventional significance the utterance has—what linguistic act is being performed by that utterance. The latter determines whether the utterance is an assertion, command, question, expression of wish, etc. Following Frege's terminology, Dummett calls the first component of meaning: sense, and the other: force. (LBM, p. 114) The necessity of drawing this distinction is by no means incumbent upon us; for instance, some paragraphs of Wittgenstein's *Philosophical Investigations* can be seen as inveighing against it. Nevertheless, as Dummett claims, without this or a similar distinction it is hard to conceive of how one may set about building a meaning theory. (WTM, p. 72) As a consequence of the failure to draw the distinction, ascribing to a speaker a grasp of the meaning of a sentence also credits him with knowing every feature of its possible use. Additional support for drawing this distinction comes from an inclination to see in the meaning of an expression something that does not change relative to whether the expression occurs in an assertoric utterance, command, wish, or question. There is something invariable in the meaning of 'tea', no matter whether this expression occurs in an assertion, e.g., 'Your tea is on the table', in an expression of wish e.g., 'Could you give me some tea?' or in a command. Once it is assumed that senses of expressions are invariant under changes of force, we face two tasks, of forming an explanation of senses and then of accounting for forces. The former is seen as independent of the latter. As to their relation, the ideal envisioned in Dummett's methodology for possible meaning theories is that the part of meaning theory that accounts in terms of some notion or other for senses of expressions of a given language should be liable to supplementation by principles linking a grasp of senses with a knowledge of possible forces. Accordingly, from a speaker's knowledge of the account of senses and the principles mentioned it should be possible to derive his understanding of a given utterance: its sense and the kind of the linguistic act performed. Now, concentrating only upon the assertions of a language, the sense of an expression is assumed to be the component of its meaning that goes into the determination of truth or otherwise of any sentence in which that expression occurs. (LBM, p. 114) This formulation is thoroughly Fregean, though the concept of truth need not necessarily be identified with the truth of classical two-valued semantics. As one task that Dummett assigns to semantic theories is to show, given the assumed concept of truth and semantic values, how the truth (or otherwise) of a sentence derives from its composition and the semantic values of its constituents, there is a certain affinity between the objective of the semantic theory and the definition of sense: the concept of sense for a given language will draw on the semantic theory for that language, in particular on its concept of truth and the assumed mode of determination of truth or otherwise of sentences of that language. It is this link that lies behind Dummett's case against the validity of classical logic in mathematical reasonings, and for this reason we will further be concerned almost exclusively with senses, or possible theories of sense. As to the other component of meaning, force, it is responsible for the ways expressions of a given language are used. If we think of a meaning theory as encompassing all modes of knowledge of language, that part of it that deals with force will put the theory in touch with the actual employment of sentences. (LBM, p. 115) The mere understanding of senses of expressions puts one only half way to a mastery of language, as such a speaker can perhaps grasp the content of utterances, but is unable to understand their significance in given circumstances. Additionally, as Dummett hopes to get the concept of truth from the actual employment of assertions, the burden of distilling that concept falls on investigations pertaining to forces of utterances (*ibid.*).

A natural question for someone aiming at a systematic account of sense is, What are the smallest linguistic units to which senses attach? A commonsense answer to this query, shared also by pre-Fregean philosophy, points to words as the basic units of discourse. The implausibility of this choice, however, reveals itself if one reflects on the linguistic objects by means of which we make 'moves' in our use of language. Another helpful reflection is on how expressions are explained, for instance, what form such explanations as given in dictionary or encyclopedia entries have. With some exceptions, and these can be viewed as abbreviated forms of sentences, our basic moves in discourse are effected by sentences. The explanations of senses, which people or dictionaries give, usually indicate what role a given word may play in sentences, and exhibit by examples how the word occurs in a typical sentence. This matches well Frege's famous contextuality principle that requires the meaning of an expression to be considered in the context of a sentence, and never in isolation. Thus, the preferred approach to senses should start with senses of sentences, and then derive, in a systematic way, senses of those expressions that are less complex than sentences, a guide being that the sense of an expression is its contribution to the sense of any sentence in which it may occur. Now, this priority of sentences seems to conflict with the observation that for a substantial part of sentences we derive our understanding of them (i.e., grasp of their senses) from our understanding of their constituent expressions and the way they are put together. This ability appears to be necessary for our mastering of those expressions that we have never heard or read before. This observation underlies the principle of the compositionality of senses, namely, that the sense of a complex expression is compounded from the senses of its constituents. (LBM, p. 137) However, it seems that compositionality taken together with priority of sentences involves one in a circular explanation. In order to understand a sentence, we need to grasp the senses of its constituents, with this requiring us however to understand sentences, presumably of any complexity whatsoever, in which they may occur. But the understanding of those latter sentences presupposes our familiarity with the senses of their component expressions, and that again is based on our grasp of the senses of sentences in which they may occur. If this were so, the grasp of any given expression would demand the command of the entire language. Thus, to block this consequence, one must put a limit on the complexity of sentences whose senses must be grasped for the understanding of a given expression. Accordingly, to grasp the sense of a given sentence one need only be aware of the senses of sentences of lower complexity than the one considered and such that its component expressions occur in them. The imposition of such a limit has two effects. On the one hand, understanding the sentence presuppose familiarity with some proper part of the language. On the other, understanding a language is not an all-or-nothing matter, since the grasp of a sentence does not presuppose understanding of sentences of every possible complexity. Given we had in natural languages a viable concept of the complexity of sentences, the idea is that there is a ranking of sentences that proceeds in accord with their complexity and such that a grasp of one presupposes understanding of the others. (EI, p. 368) As we shall see, Dummett puts a strong emphasis on this requirement, as he believes that its satisfaction precludes meanings of the language from being holistic, where by holism he understands the doctrine that the grasp of the sense of a sentence entails understanding of the whole language.

3.3. *Sense and Semantic Theory*

In order to come to terms with the somewhat nebulous idea that the meaning theory for a given language should be erected on the basis of a semantic theory for that language, we need to consider Dummett's conditions concerning senses. They can be viewed as a generalization of Frege's principles, with a caveat however that where Frege takes a definite stance of identifying references of proper names with their bearers and references of sentences with their truth-values, Dummett remains neutral in speaking of 'semantic values'. We have the following five principles:

i. To give the sense of an expression is to give a complete characterization of a piece of knowledge that the speakers have concerning it.

ii. Given how the world is, sense determines semantic value.

iii. Given principle (i), nothing belongs to sense save what is required to determine semantic value (...).

iv. The sense of a complex expression is compounded out of the senses of its constituents.

v. An expression has sense only in the context of a sentence. (LBM, p. 137)

The first principle expresses the discussed methodological postulate of approaching the queries concerning the sense of an expression as pertaining to the knowledge a speaker needs to have in order to grasp that sense. As to the second condition, it reflects an intuitive idea that there are exactly two factors that determine what semantic value the expression has: the sense of this expression and the way the world is. If we give to (ii) a more epistemic reading, which we obviously can since sense is correlative to knowledge, it amounts to the claim that the speaker's grasp of the sense of the expression should help him to determine what the semantic value of that expression is. This formulation is intentionally vague, as the nature of that determination is dependent on the meaning theory and its underlying semantic theory, an important difference being the one between meaning theories that make the speakers effectively able to determine semantic values, and meaning theories that cannot guarantee this. The third principle delineates the knowledge that goes into someone's grasp of sense, relegating from sense all information concerning an expression which is irrelevant for the determination of the semantic value of the expression. The next condition is the requirement of the compositionality of senses, namely, that the sense of a composite expression should be systematically read from the senses of its components and the way they are put together. On Dummett's construal there is more to this principle, however. It is essential to the sense of a composite expression that it be compounded in the way it is and from those constituents that it actually has. For, to grasp the sense of a compound expression, one needs to be aware of its composition and the senses of its constituents. If one arrives at the opinion that some other expression, of a different composition, has the same sense as the original one, he must be aware that the sense in question is capable of being conveyed by an expression whose syntactic composition is that of the original one. (LBM, p. 137) Finally, the last condition is a version of Frege's contextuality principle, though in contrast to Frege who stated it for an undifferentiated concept of meaning, it here concerns senses only.

Now, the account of senses is related to the semantic theory *via* principles (ii), (iv) and (v). According to condition (v), it is essential for expressions to figure in sentences: the senses of expressions combine together so as to yield the sense of the resulting sentences. To this (iv) adds that the senses of constituents and the way they are put together determine uniquely the sense of the sentence. Clearly, on the level of semantic values the exact analogue of (iv) is the claim that the semantic value of the compound expression is determined by the semantic values of its constituents and the way the constituents are put together. The two levels are bridged by principle (ii): the sense of each constituent expression determines its semantic value, and similarly the sense of the resulting sentence determines its semantic value. Now, could it be possible that there is a disparity between the two levels, so to speak? That would mean that the structure of senses that combine to yield the sense of the compound expression diverges from the structure of semantic values that go into the determination of the semantic values of that expression. This is, however, precluded by principle (iv) that intervenes to assure that the sense of a given expression must be thought of as possessed by the expression of a certain syntactic structure. If you recognize that a given sense is identical with that of this expression, you must know that it is conveyable by the expression of exactly this syntactic composition. Thus, semantic theory yields the following contributions to meaning theory. First, by choosing semantic values attributable to each category of expressions, it gives substance to the idea that knowledge of the sense of an expression determines its semantic value, as it delineates what this determination is and indirectly how it is effected. Secondly, it tells how to think of the sense of an expression as being compounded from senses of the constituent expressions. Finally, granted that it is the task of an account of senses to clarify how the sense of a sentence is related to its truth or otherwise, the answer must draw on that part of semantic theory that shows how the semantic value of the sentence is linked to its being true or otherwise.

Let us now turn to the language of classical mathematics. To inquire about the semantic theory for classical mathematics is to ask what semantic theory can justify the forms of reasoning applied in classical mathematics, that is, forms of reasoning of classical logic. The general answer is that the valuation system must have the structure of a Boolean algebra. In the simplified case the algebra turns into the familiar two-valued semantics, a valuation system being identified with the two-element algebra $\mathbf{Z}_2$. The two elements are interpreted as objective truth-values, truth and falsity, one of which determinately attaches to each sentence of the language of classical mathematics. The generalization to arbitrary Boolean algebras is needed if a mathematical theory is taken not to have one intended model, but as requiring a number of equally admissible models, it being assumed that every sentence is either true or false in each model. Accordingly, the evaluation relies on assigning to each sentence a set of models in which it is true. To put it in Dummett's terminology, statement values are identified with sets of admissible models. The difference between two-valued semantics and an arbitrary Boolean algebra as a valuation system corresponds to the divergence between two views on whether a sentence shown independent from the assumed axioms of a mathematical theory can have absolute truth values. Proponents of the first option believe that in doing mathematics we think of a single intended model of a given theory, the model that we somehow intuitively see. Results showing independence of some statements of that theory from its axioms are then interpreted as a failure of the axioms to capture our intuitions, or the way we think of the intended model—on this construal such a situation reveals a failure of our axiomatization and points to a limitation of the formal method. The alternative school concedes that it may well happen that a theory does not have an intended model but only a set of equally admissible models and accordingly denies that independence results testify to the axiomatization being inadequate in the above sense. (EI, p. 370) However, from the point of view of the intuitionist, both these variations share the same feature at which the nub of the intuitionistic critique is directed: a statement value, no matter whether it is a set of admissible models or simply a truth-value, determinately attaches to each sentence, and this is so regardless of whether a competent speaker can recognize what statement value the sentence has or not. Thus, in discussing these objections, we will follow Dummett in the concentration on two-valued semantics.

4. THE LANGUAGE-LEARNING ARGUMENT

4.1. *Knowledge of Truth Conditions*

A conundrum appears to follow from combining the principles concerning senses with classical semantics. According to principle (ii), given how the world is, the sense of a sentence should determine its semantic value. This formulation leaves unspecified how to understand the nature of that determination. Now, in the two-valued semantics, semantic values of sentences are identified with truth or falsity and semantic values of singular terms—with their bearers, the consequence being that given how the world is, the sense of a sentence should determine its truth or falsity, and the sense of a singular term—its bearer. That means that generally the determination of semantic values through senses stops being humanly realizable. We understand many sentences whose truth or falsity remain hidden to us, including also those whose truth-values we will perhaps never be able to discover. A similar observation holds true of names: we appear to understand a name although it may happen that we are not able to recognize its bearer. Facts like these require the relaxation of the requirement on what the determination of semantic values by senses is, with the concession that it may not be effective. Once this is done, the determination loses much of its epistemic character and, as all this turns rather foggy, a recommendable solution would be perhaps to give up principle (ii). This, however, would mean that two-valued semantics is not at all a suitable candidate for the foundation of a meaning theory. Thus, one is forced to acquiesce in the rather spurious idea that by knowing the sense of a sentence, a person somehow knows what it is for that sentence to be true. This is expressed by the familiar slogan that the understanding of a sentence consists in knowing its truth condition, the underlying assumption being that truth (and its twin notion—falsity) objectively attach to the sentence. Now, in accord with the requirement of linking meaning to the knowledge of a competent speaker, Dummett asks what is involved in ascribing to a person the knowledge of the truth condition of a sentence. In many cases we are ready to ascribe to someone the knowledge of the truth condition of a sentence on the grounds of his ability to state it in other words. Presumably, it is even not realistic to require of someone who grasps the truth condition of the sentence 'The president-elect is being sworn on the steps of the Capitol' anything more than an acceptable translation that explains the notion of swearing-in together with the institutions of presidency and Congress. Nevertheless, as it is impossible to state in a given language the sense of each expression without falling prey to a vicious circle, there must be some part of language knowledge about which is explained differently. Since expressions belonging to this part of language are believed to be learned by observing how others use them, the requirement emerges that the grasp of truth conditions of sentences belonging to this part of language should be manifested in use. Thus, we have two modes of knowledge of truth conditions, which may be called after Sundholm (1986, p. 481) *statable* and *non-statable* knowledge. Statable knowledge is ascribed to a person merely on the grounds of his ability to express it in other words, whereas of someone possessing *non-statable* knowledge it is required that this knowledge be manifested in use. Although it is notoriously unclear whether or not for a given sentence its understanding falls under the first or the second category, the picture is intuitively supported by reflecting on the learning of the mother tongue, where we are inclined to make a similar distinction between a part of language that is learned by observing in what circumstances utterances are made and the other part one becomes acquainted with by means of purely verbal explanations. Now, as we face the problem of explaining how one may attribute to a person knowledge of the truth condition of a sentence for which the decision procedure is not known, the temptation is to contend that the understanding of such sentences is statable knowledge. This response, however, runs counter to the observation that undecidability is not brought to language as a result of some new expressions being introduced by purely verbal explanations of their meaning. That is, if a language does not contain yet-undecided statements, i.e., statements for which decision procedures are unknown, the mere addition of new expressions that are expressible by means of the old vocabulary will not introduce any 'new' yet-undecided statements. Consequently, if the language permits the formation of sentences for which decision procedures are unknown, then the grasp of the truth conditions of some sentence of this sort must be non-statable knowledge. This means that the proponent of two-valued semantics owes us an explanation of how knowledge of the truth condition of a yet-undecided sentence is manifested.[48] Before we come to this problem, let us however explain the easier one, namely how to think of *manifesting* the knowledge of the truth condition of a decidable statement, where 'decidable' means that a decision procedure for that statement is known. In Dummett's somehow elaborate terminology, to manifest fully the truth condition of such a sentence an individual must display: (1) mastery of the relevant decision procedure, that is, the procedure which allows one to get into a position in which he can recognize the truth condition as obtaining, if it obtains, and (2) verbal behavior by which the individual acknowledges the condition as obtaining if it obtains (or as failing to obtain if it fails to do so). To take an example, given that a student is being examined on whether or not he grasps the truth condition of the sentence: '31 is a prime number', he should in accord with the above recipe make his teacher observe his failed attempts to divide 31 by $2, 3, 4, \ldots, 30$ and then make him notice that he accepts the number as prime.

Now, if we turn to a yet-undecided arithmetical statement of which it is assumed that knowledge of its truth condition is not merely statable, for obvious reasons the above sort of explanation does not work. As Dummett puts it:

Since the sentence is, by hypothesis, effectively undecidable, the condition which must, in general, obtain for it to be true is not one which we are capable of recognizing whenever it obtains, or of getting ourselves in a position to do so. Hence any behavior which displays a capacity for acknowledging the sentence as being true in all cases in which the condition for its truth can be recognized as obtaining will fall short of being a full manifestation of the knowledge of the condition for its truth: it shows only that the condition can be recognized in certain cases, not that we have a grasp of what, in general, it is for that condition to obtain even in those cases where we are incapable of recognizing that it does. It is, in fact, plain that the knowledge which is ascribed to one who is said to understand the sentence is knowledge which transcends the capacity to manifest that knowledge by the way in which the sentence is used. (PBIL, p. 319)

Let us attempt a little exegesis of this fragment. Note first that what Dummett dubs 'undecidable statement', i.e., statement for which no decision procedure is known at a given time, bears here the more accurate name of 'yet-undecided statement'.[49] What is then involved in saying that, in the case of a yet-undecided statement, a person cannot recognize or put himself in a position to recognize its truth condition as obtaining, if it obtains? As the hypothesis is that the statement remains undecided, we cannot manifest the grasp of its truth condition by revealing mastery of the decision procedure. Perhaps one may invent some other way of fully manifesting the truth condition, but it is hard to say what it may be. What then is this strange entity, the truth condition, that supposedly escapes our recognition? Taking it literally, we may think of it as some 'possible' structure in the mathematical realm such that given it holds, our sentence is true. However, Dummett is not concerned with the question of what the truth condition is, but rather what is involved in saying that someone knows it. We ascribe to a person the knowledge of circumstances under which the sentence is true; however, as the person neither knows a truth-value of that sentence, nor even a procedure that would show whether or not it is true, he is not able to recognize whether that condition obtains, whereas in accord with the two-valued semantics, the truth condition either obtains or not. In short, the condition obtains or does not obtain, the person is said to know that condition, yet he is incapable of recognizing whether it obtains, if it does. There is an inclination to diagnose this predicament by saying that the ascription of knowledge has gone awry. But is this inclination correct?

At first glance, it looks rather natural that whenever we credit someone with the grasp of some condition, we expect that person to be able to recognize whether the condition obtains, if it does. For instance, given that the knowledge of the conditions of the application of a color-word, say, 'yellow' is ascribed to an individual, that person is expected to recognize, whenever presented in appropriate circumstances with a yellow object, that it is indeed yellow, where the circumstances include such factors as the person being close enough to the object examined or the proper lighting. This makes clear, however, that the ability to recognize the condition as obtaining is conditional: given that the circumstances are appropriate, the person should recognize that the object is yellow, if it is yellow. What then about circumstances that have not obtained throughout a relevant period, but are nevertheless necessary for recognizing whether a condition obtains? Could we cogently ascribe the knowledge of the condition to a person who, due to the lack of the proper circumstances, has not been able to recognize the condition as obtaining, though it has obtained? There is no clear-cut answer to this query but we may try the following story. Suppose that there are certain spectacles that make people who wear them blind to some relatively rare colors. For instance, one spectacle of this sort prevents a person from differentiating between sea-green and plain green, another blurs the difference between purple and red, and so on. Once they are removed from the person's eyes, however, he immediately recovers normal vision. Suppose now that there is a person of whom we know that he has been wearing several pairs of such glasses from birth. He has been taught, moreover, that there are more colors than he actually perceives, so he even has in his vocabulary the names for these colors. That is, he knows that once he removes the 'green' glasses, he should be able to perceive a new color, called the 'sea-green' and if he removes the red glasses, the new color he will see is named 'purple' and so on. The question is: Could we credit him before he takes off the spectacles with knowing the condition of the application of the word, say, 'purple'? Evidently, in the circumstances considered, he is not able correctly to apply this word to purple objects he is confronted with. Nevertheless, judging from the fact that he correctly applies color-words to objects whose colors the glasses do not interfere with and the fact that he knows the workings of the glasses, there is an inclination to say that the person knows the condition of the application of 'purple'. For, one may argue, if he only took off the glasses, he would recognize whether or not the condition obtains. Moreover, we may develop our story further, so as to include our earlier observations about his taking off other glasses of this sort and soon afterwards correctly applying, say, the word 'sea-green' to sea-green objects. Finally, we may take his own words for his knowing the condition of the application of 'purple' since he is able to state that 'purple' applies to objects whose colors he is not acquainted with as long as he wears red spectacles and purple is the new color he will perceive as soon as he takes off the glasses. Again, judging from how he behaved after taking off other glasses of this sort, we may give much credibility to his words, though they merely *state* what the condition is. Of course, there are too many idealizations involved in our story to claim that it represents our use of color-words. It points only to the cogency of the ascription of the knowledge of a condition to a person who, due to the circumstances he has been in, has not been able to recognize that the condition obtains, even though it has obtained in his vicinity all along.

Is there any affinity between the example and the issue of ascribing to someone the knowledge of the truth condition of a yet-undecided statement? To be tempted into this similarity we need first to think of the truth condition that obtains as somehow presented to a person considered, either all the time, or (perhaps better) whenever that person contemplates the content of the statement. Indeed, Dummett writes as if the truth condition of a yet-undecided statement were present, so to speak, before the eyes of a person who contemplates that statement, as if he confronted it, but could not recognize it as obtaining, if it obtained. Clearly, an analogue of the circumstances necessary for recognizing whether or not the object has a certain color is provided by knowledge of a procedure that permits one to decide whether or not the sentence is true. Consequently, as long as that knowledge is lacking, the condition is not recognized as obtaining, given that it obtains. Finally, the ascription of knowledge of the truth condition of a yet-undecided statement may be made on the grounds of that person's manifesting knowledge of the truth conditions of decidable statements of a similar syntactic structure. Again, his ability to recognize as obtaining the truth condition of a statement for which the decision procedure has only recently been found may testify to his grasp of the truth condition before the procedure was discovered. Given this, we may even take his own word for his knowledge of the truth condition of a yet-undecided statement, as he has proved himself trustworthy in earlier cases. Thus, even a mere verbal explanation of what the truth condition is may count as a manifestation of knowledge of it, given that it is made against the background of the relevant information. All this means, thus, that there are various pieces of information on which the ascription of understanding the condition draws, and only rather simple conditions can be subsumed under the model of knowledge-ascription that Dummett works with and which requires anyone said to know the condition to be able to recognize it as obtaining, if it does obtain. There are cases where this model works quite well, for instance cases related to the understanding of occasion-sentences. Given that the person understands such a sentence, he is able, for any state of affairs, to decide whether he may assent to the sentence, or should rather dissent from it. There does not seem to be any reason, however, why this simple model should provide a standard for understanding other sorts of sentences, for instance mathematical ones.

Thus, the meaning theory framed in terms of truth conditions, with the implication of ascribing knowledge of the truth conditions of yet-undecided statements, does not lead to the difficulties Dummett points out. Nevertheless, there is a certain oddity in such theories, revealed in the question of what these things, the truth conditions of mathematical statements, are. If we construe them analogously to the truth conditions of, say, statements relating to material objects, they will be viewed as mathematical facts or states of affairs located in some mathematical realm that the person somehow confronts without being always able to recognize whether they obtain, if they do. Perhaps these fictitious entities are a part of the package deal of purchasing the literal construal of the notion of grasping truth conditions, reinforced further by the semantic requirement that the sentence is determinately either true or false. If so, the proponent of a truth conditional meaning had better suggest some other construal of the notion of grasping truth conditions that avoids the above pitfall. A natural move, though clearly *ad hoc*, is to identify grasping the truth condition of the statement with the ability to decide, whenever presented with a mathematical construction, whether or not it is a proof of this statement, where the proof appeals to the classical forms of reasoning only. If we think of this in terms of a formal mathematical theory, given that the provability is decidable, there is no problem with ascribing to a person the ability to recognize whether or not the condition holds. A purist may correctly object to this proposal by asking why this should be called 'grasping truth conditions', as it is clearly an initial step towards characterizing meaning in terms of provability. A partial answer would point to the fact that according to this proposal meaning of sentences is given in respect to *classical* provability, that is by means of proofs that are subject to justification by two-valued semantics, or in a general case, by a valuation system that constitutes a Boolean algebra. Whether this proposition is cogent, and satisfies Dummett's requirements for a tenable meaning theory, is a debatable question to be returned to later. It is of critical importance for Dummett's argument for intuitionism, however, for even if it turns out that the truth-conditional meaning theory is to be replaced by a theory framed in terms of provability, we will still face the problem of why it should be *intuitionistic* provability.

4.2. *The Ingredient of Meaning That Transcends Use*

While arguing against Dummett that in some cases the inability to recognize a condition as obtaining when it obtains does not tell against a person's knowledge of this condition, we have left a question unanswered. It amounts to asking what the significance of the requirement of full manifestability is and how to understand this notion. The requirement is perhaps best approached through Wittgenstein's identification of meaning with use, as Dummett's objection is that knowledge of truth conditions presupposes having some 'ingredient of meaning' that transcends any possible use or that the knowledge involved can never be *fully* manifested. Why, however should one stick to Wittgenstein's *dictum*? Dummett proposes two ways of defending the claim: *via* communication of meanings and *via* a picture of learning a language. The key requirement for linguistic communication that he accepts is:

(...) if two individuals agree completely about the use to be made of the statement, then they agree about its meaning. (PBIL, p. 216)

If this were not so, we could hardly communicate in language, since then to understand the other speaker's utterance it would be necessary to grasp some factor that is not revealed, and hence cannot be observed, in the way that the other person uses language. Somewhat fictitiously, even if two persons made a thorough examination of the ways in which they use a given sentence and found out that they were prepared to assert it in the same circumstances, and also to withdraw its assertion in the same circumstances, in short, that their uses of that sentence were the same, they would still be perplexed as to whether they meant it the same way. And their perplexity would not stem from the fact that the examination was not thorough enough, but from the possibility that each attaches to that statement some hidden element of meaning that is not revealed in such an investigation, no matter how long or detailed the procedure. The supposition of such a hidden element of meaning leads to the consequence that meaning is not communicable. If two individuals agree on even the tiniest aspects of the possible use of some statements, this does not count as evidence that they understand each other. Nevertheless, the spurious part of this reasoning lies in the fact that it turns on a terminological issue. On the one hand, if the notion of meaning is construed as a property correlative to our understanding, and possessed by expressions in virtue of their belonging to social language, then an understanding of an expression that requires the apprehension of some hidden factor in addition to grasping the meaning is not conveyable, though meaning is communicable *ex definitione*. On the other hand, if we neglect the social character of meaning and place meanings in the speakers' head, then we are in for deriving a consequence that meaning is conveyed only partially, as there may always be different factors that go into meanings that individuals ascribe to the same expression.

Be that as it may, we do not need to be troubled with this terminological issue, as the second of Dummett's arguments is rather neutral in respect to it. The starting point is the following picture of learning the language of mathematics:

When we learn a mathematical notation, or mathematical expressions, or, more generally the language of a mathematical theory, what we learn to do is to make use of the statements of that language: we learn when they may be established by computation, and how to carry out the relevant computations, we learn from what they may be inferred and what may be inferred from them, that is what role they play in mathematical proofs and how they can be applied in extra-mathematical contexts, and perhaps we learn also what plausible arguments can render them probable. (PBIL, p. 217)

Those various aspects of uses of the statement are all that can be revealed to a person who learns that language. Accordingly, they provide the only evidence for whether or not a person understands a mathematical statement. A supposition that there is something more to grasping the meaning of the statement, something that is not revealed in any possible use made of that statement, lands us in a position in which linguistic communication is illusory. Assume for instance that a person has been shown various aspects of the uses of some statements, and now uses these statements in the way he was taught. Even if you are not skeptical as to whether on some future occasion his uses of those statements will diverge from those taught to him, on the supposition that there is some additional factor, over and above abilities to use these statements, that is necessary for their understanding, you conclude that there is no evidence that the person understands the statements that he was expected to learn. In this view teaching or examining thus has little to do with understanding, as it is impossible to teach anyone to understand, say, a mathematical notation, in the way a teacher does. Consequently, people can only be taught to imitate the verbal behavior of others and be examined as to how well they imitate it. This would be all that teaching or examining could accomplish. Given the social character of mathematics and other sciences, this conclusion seems, however, absurd. We thus conclude that understanding the expression, or the grasp of its meaning, should be capable of being fully revealed in its use.[50]

Now, our central problem is whether equating understanding of the statement with the grasp of its truth condition commits one to maintaining that there is an ingredient of meaning that 'transcends use', or that understanding of the statement is never 'fully manifested'. Dummett explicitly suggests that the latter consequence obtains in the case of yet-undecided statements, i.e., statements for which no decision procedure is known at present. As he puts it, it is not that the grasp of its truth condition cannot be manifested; rather, the emphasis is that this is not a *full* manifestation:

(...) [A]ny behavior which displays a capacity for acknowledging the sentence as being true in all cases in which the condition for its truth can be recognized as obtaining will fall short of being a full manifestation of the knowledge of the condition for its truth: it shows only that the condition can be recognized in certain cases, not that we have a grasp of what, in general, it is for that condition to obtain even in those cases when we are incapable of recognizing that it does. (PBIL, p. 225)

So, the objection is that if a statement is true, the condition invariably obtains; yet despite our supposedly knowing what it is for that statement to be true, we do not recognize the condition as obtaining at least as long as it remains undecided, and *this* supposedly testifies to this knowledge not being fully manifestable. It is implied then that if it happens that sometime later a person proves the statement that was previously undecided, then his carrying out of the proof shows only that on a *certain* occasion did he know that the condition obtains, and this fact supposedly does not count as full manifestation of the knowledge of the truth condition. Accordingly, the full manifestation of the grasp of the truth condition must consist in the ability to recognize, on *any* occasion, that it holds, if it does so. This, however, appears implausible, at least if one compares knowledge of truth conditions with knowledge of other conditions, say, of the application of a color-word. Given that a person was unable in some circumstances to correctly apply the word 'green', but later does it very well, we may, perhaps drawing on additional information, say that this not only manifests, but also *fully* manifests his knowledge of the condition of the application of 'green'. There is no reason for believing that his correctly applying that word from some later time on must fall short of being a *full* manifestation of the relevant knowledge. Of course, a bit of paradox still lurks in this comparison. Since in both cases there must be an explanation of why the condition was not recognized earlier, the missing knowledge of the proof will be considered somehow on a par with circumstances like improper lighting or wearing glasses that disturb the vision.

However, what about truly undecidable statements, or at least statements of which it is not known for a sufficiently long period whether or not they are true? In that case, the truth-conditional meaning theory ascribes to speakers some knowledge that *ex hypothesis* cannot be fully manifested throughout their lives, where the last notion is explained as usual as involving mastery of the relevant decision procedures. At first glance, to ascribe to someone the knowledge of a condition that he has had no occasion to display throughout the whole his life seems implausible. This is like ascribing to someone who has lived all his life in an area where there were no pink objects the ability to recognize the color pink on the grounds that had he lived somewhere else, he would have correctly applied this word. A plausible retort may, however, be that this is how the ascription of knowledge works. Since manifestation of knowledge is conditional on circumstances, there will be such extreme cases. And, moreover, we will face essentially the same problem even if meaning is framed in terms of provability, as Dummett recommends. We will then be ascribing to a person the knowledge of what proof may render a given statement true; but, as no proof of that statement may be known, the person may have no possibility of recognizing it, and hence, manifesting fully the knowledge of its provability-condition.

Dummett's argument must thus exploit a discrepancy between, on the one hand, statements of which in the period of carrying out the meaning-theoretical investigations it is known how to prove or disprove them, and, on the other, statements that remain undecided in that period. The discrepancy is revealed in the alleged incorrectness of ascribing to someone knowledge of the truth conditions of the latter sentences, whereas there is nothing intricate in the ascription of the grasp of truth conditions of the former. However, for the argument to work, it must be seen as implying that there is a problem with knowledge of the truth condition of a statement whose proof was not known from the beginning of our investigation, but was found somewhat later: Someone's mastery of carrying out the proof is said not fully to manifest his grasp of the truth condition since this person was unable to recognize the truth condition before the procedure was found. That, however, verges on absurdity, as it makes the notion of full manifestation relative to the period of the meaning-theoretical investigations. Someone who displays a mastery of the decision procedure, but was not earlier able to do so because the procedure was not known, is said not to be fully manifesting the grasp of the truth condition, but another person, who was not tested earlier and now shows mastery is seen as fully manifesting that knowledge.

Let us turn finally to the lingering question of why the full manifestation of the grasp of the truth condition of a statement must involve nothing but the mastery of the relevant decision procedure. If we turn to first-order arithmetic, yet-undecided statements are typically expressed by sentences with the general quantifier as the principal operator, that is, $\forall_x A(x)$, where for any given term t it can be decided whether or not $A(t)$ holds. Why, then, can we not content ourselves with saying that a grasp of the truth condition of a yet-undecided statement expressed by a sentence of the form $\forall_x A(x)$ is fully manifested in the way a person recognizes whether or not $A(t)$ holds for some t, taken together with his ability to use the general quantifier as shown in his mastery of recognizing as true or false decidable statements of the form $\forall_x F(x)$? It may be argued, moreover, that a rationale for this move is provided by the principle of the compositionality of meaning. Since the meaning of a statement is determined by the meanings of its constituents and the way they are put together, a similar idea carries over to the ascription of knowledge: given that we credit a person with understanding the constituents of the sentence as he is observed to use them correctly in sentences expressing decidable statements, and is seen to know the syntactical composition of that sentence, we should rather expect him to grasp the meaning of this sentence. Thus, at least as long as it is not claimed that an expression may change its meaning depending on whether or not it occurs in a sentence expressing a decidable statement, there does not seem to be any reason to hold that the grasp of meaning cannot be fully manifested in such a piecemeal way. But to hold that, say, 'odd number' may mean something different in Goldbach's conjecture and in some decidable statement is a curious position, to say the least.

It is rather surprising that Dummett, for whom the principle of the compositionality of meaning is so dear, has never envisaged the possibility that knowledge of the truth condition of a sentence can be construed as a grasp of the meaning of its constituents (again explained in terms of their impact on the truth conditions of other sentences), and that full manifestation of the knowledge of the truth condition may also consist in a display of the understanding of constituent expressions (i.e., in showing mastery in deciding whether or not statements in which the relevant expressions figure are true). Perhaps he abstains from making this move as he fears that this will lead to a kind of holism such that "(...) no sentence of a language can be fully understood unless the entire language is understood." (EI, p. 366) More precisely, one may feel uncomfortable with the fact that such an approach prohibits ordering sentences with respect to their complexity so that, in order to know the meaning of some sentence, it is not necessary to understand sentences of an arbitrarily high complexity. Clearly, on investigating whether a speaker understands the universal quantifier, one can choose any decidable statement and observe how the speaker carries out the relevant decision-procedure. Nevertheless, there does not seem to be any obstacle against imposing such a restriction to the effect that, say, evidence for someone's understanding of the general quantifier can be gathered by observing how he handles statements of a lesser complexity than a given one.

Thus, it seems that Dummett is forced into this narrow concept of manifestability by holding on to some methodological principle which demands that each term of a meaning theory should correspond one-to-one to some specifiable, distinguishable sort of abilities constituting understanding of a language. (EI, p. 377) This demand applies also to the crucial term: 'knowledge of truth condition'. Our proposal to relax the standards of what constitutes full manifestation of grasping the meaning of a yet-undecided statement is in conflict with this methodological principle, since an outcome of our proposal is that if a speaker understands a yet-undecided statement, it is hard to say what other sentences he should understand and what decision-procedures he should be capable of carrying out. Why, however, should we adhere to Dummett's methodological principle? As it stands, the principle is hardly observed in the natural sciences. Thus, Dummett's demand, as explained above, is merely the methodological principle of meaning theories. The principle comes from assuming that a meaning theory should not be compared, say, to physics; it should be neither a science that postulates posits nor a science having a theoretical vocabulary. Instead, it ought to account, in common-sense terms, for what it is to understand a language. If this diagnosis is correct, it turns out that the argument for intuitionism hinges heavily upon a methodological principle which, taken generally, is if not false, then at least highly dubious. In addition, justification of the principle invokes the claim of the specific character of meaning theories. There is at present, however, no fully worked-out theory of this kind, and one cannot ensure that there will be any in the future. It is thus hard to see whether the methodological demand can be satisfied and whether such theories may have the specific character mentioned. Thus, there does not seem to be any reason why one should hold to Dummett's methodological principles, and the resulting narrow concept of the full manifestation.

We have been investigating whether the truth conditional account of meaning commits its proponent to a belief in a hidden ingredient of meaning, which can never be fully manifested. In fact, however, no such ingredient, or as Dummett sometimes puts it, 'mental content', is presupposed by this account. There is a certain oddity in ascribing knowledge of truth conditions of yet-undecided statements to people. But the oddity disappears once it is noticed that a manifestation of that knowledge is conditional on circumstances. Moreover, Dummmett's argument draws on a rather peculiar construal of the notion of *full* manifestability, its peculiarity stemming from requiring that it is nothing save the ability to display a relevant decision procedure on any occasion. This leads to absurd consequences in the case of a once-undecided statement being proved (or disproved). Since this construal of full manifestability is not motivated by the desire to avoid holistic conclusions, there is an inclination to locate its source in the methodological ideal of the meaning theory, which, however, can hardly be justified. Thus, there is no obstacle to understanding the notion of manifestability much more broadly, the consequence being that knowledge of the truth condition of a yet-undecided statement is fully manifestable.

4.3. *Why Intuitionistic Provability?—Holism to the Rescue*

Although Dummett's objections to the truth-conditional account of meaning are rather off target, a certain awkwardness lurking in the notion of truth condition as applied to mathematical statements, nevertheless provides a rationale for searching for other accounts of meaning. The awkwardness is due to the idea that if a statement is true, its truth condition obtains, which invites in turn the question whether we are 'presented' with the truth condition, and in case we are, why we might fail to recognize it as obtaining. The proposal Dummett advocates is to choose provability as the central notion of meaning theory, with the consequent claim that someone grasps the meaning of a statement if he knows what counts as a proof of it. Does this, however, offer any advantage over the truth-conditional account? If we turn to formal theories, as long as provability is decidable, which is the case in both formalized intuitionistic mathematics and formalized classical mathematics, it may be safely assumed that once presented with a formal proof of the statement, a person will be able to put himself into a position in which he recognizes that it is a proof of the statement in question. Clearly, there are many statements whose formal proofs are not known, and this raises the question of how their understanding should be represented. To be in accord with Dummett, we need to require that the person who understands such a statement should be able to recognize its proof when presented with it—this alone counts as the full manifestation of his grasp of its meaning. This means, however, that as long as the statement remains neither proved nor disproved, the individual cannot fully manifest his grasp of its provability condition, as he cannot manifest its truth condition. Accordingly, yet-undecided statements still pose a problem—so far the appeal to the provability condition does not offer grounds for preferring intuitionistic provability as the key notion of meaning theory.

Why then should it be intuitionistic proof and not that of classical mathematics, or any other sort of proof as long as the forms of reasoning applied are validated by a semantic theory? Given that the meaning of a statement is framed in terms of provability, it is said that understanding it, that is, grasping its provability condition, involves the knowledge of what a proof of that statement may be. That is, what is required is a *general conception* of what constitutes a proof of the statement and that needs to be derived from the composition of the statement and, if the statement is compound, our knowledge of what are possible proofs of its sentential constituents. In other words, for a compound statement there should be a uniform way of reading off its provability condition from both the provability conditions of its constituents and the principal logical operator of that statement. Now, if we concentrate on the classical disjunctive sentence, the above requirement is violated, for we have not such a general conception of how the proof of $\alpha \vee \beta$ may depend on proofs of the disjuncts. That is, given that there is a proof of a disjunct, it may be transformed into the proof of the disjunction, but the conclusion may be derived as well from $\neg\neg(\alpha \vee \beta)$ *via* the rule of double negation elimination. In the latter case it is hard to conceive of a general conception of the proof of disjunction, as again there is more than one way of arriving at the double negated sentence. (EI, pp. 368–369) Moreover, the proof of the disjunction then requires understanding of sentences of higher complexity, and as it is hard to conceive a limitation on the complexity of formulae that may occur in a possible proof of the classical disjunction, this endangers Dummett's antiholistic requirements. In this respect, the intuitionistic disjunction appears to fare better since, in accord with Heyting's interpretation, the proof of the disjunctive sentence is constituted either by a proof of one disjunct, or by a proof of the other.

Where, however, does the need for a general conception of the proof of a statement come from? Clearly, this requirement goes over and above the postulate that a condition ascribed to a competent speaker should be recognizable by him whenever it obtains. The requirement stems, firstly, from the natural postulate that it should be possible to ascribe to people understanding of statements whose proofs or disproofs are not known, the claim being that the person grasps the statement's provability conditions as he reads it off from the provability conditions of its constituents and the principal logical operator of that statement. If the intuitionistic conception of proof allows for this, it will offer grounds for preferring a meaning theory framed in terms of intuitionistic provability. Secondly, there is a certain oddity in the view according to which someone is credited with the ability to recognize a formal proof of a given statement if presented with it, while on the other hand it is conceded that the person may not have a general conception of what the proof of the statement may be. The view gains support from a reflection on formal mathematics: as long as provability is a decidable relation, it can be said that for every formula of a formal theory the person is effectively able to recognize, if presented with a formal structure, whether or not it is a formal proof of this formula. One step of this procedure consists in deciding, for any string of signs that occur in the purported formal proof, whether it is a formula of the formal theory being considered. It may thus be said that ability to recognize a formal proof of the sentence involves familiarity with all formulae occurring in this proof. But, as there is no limitation on the logical complexity of formulae that may occur in a proof of a given sentence, this presupposes familiarity with formulae of indeterminately high logical complexity. Plainly, there is nothing odd about this claim as concerning formal mathematics, since what we called familiarity with a formula amounts only to the ability to recognize whether or not the string of signs is a formula of the theory considered. And clearly, someone who is said to know a formal theory needs to know how to recognize this. The view becomes, however, controversial if it is made to apply to informal mathematics as well. It amounts then to saying that it can be recognized whether or not a mathematical construction informally proves the statement, but as there is no bound on the complexity of the formulae occurring in this proof, the assumption is that one needs to know a substantial part of the language, perhaps even the entire language. More precisely, there is then no place for a ranking of sentences such that understanding of a given one requires only understanding of some sentences of lower complexity than the one considered. Thus, the second motivation for demanding that someone who is said to know the meaning of the mathematical statement must have a general conception of what may constitute its proof, stems from the anti-holistic postulate.

The objection being that meaning in terms of the classical proof leads to holism, we need to inquire whether the intuitionist is in a better position, as he holds to Heyting's clauses stating what constructions count as proofs of logically complex sentences. One reason to be skeptical about this is an appeal to procedures that transform proofs into proofs that is made in clauses relating to proofs of a conditional, a negated statement, and a statement with the general quantifier as the main operator. To concentrate on the clause for the conditional, it is said that the proof of the conditional is a procedure that, if applied to any proof of the antecedent, yields a proof of the consequent, together with a proof that the procedure does this. As the quantification is over arbitrary proofs of the antecedent, not only do we face at once the danger of holism but, more importantly, there is a tinge of the vicious circle. Thus, we need to think of the procedure not as operating on arbitrary proofs of the antecedent, but on a restricted kind of proofs, such that the complexity of formulae that may occur in them does not exceed a certain boundary determined by the structure of the antecedent. The claim that for every statement one should have a general conception of its possible proof involves the idea of knowing how that boundary is to be determined.

Even if this problem is solved, we face the next issue, namely of clarifying the notion of procedure that transforms the proofs of the antecedent into a proof of the consequent, such that understanding of the procedure does not involve understanding the whole language. To guard against this danger, Dummett suggests putting a limitation on the kind of constructive proofs to which Heyting's clauses pertain. His contention is that a distinction must be drawn between the proofs, dubbed *canonical*, that are appealed to in these clauses, and demonstrations, that is, constructively correct informal proofs that usually occur in textbooks or mathematical journals. One reason for drawing this distinction is provided by the behavior of the intuitionistic disjunction: its (canonical) proof is constituted by a proof of one or a proof of the other disjunct, whereas on the other hand a disjunctive statement may occur at a line of a demonstration merely on the ground of there being an effective method of obtaining one or the other proof. More importantly, the distinction is necessary to avoid the emptiness of clauses for proofs of the conditional and the statement in which the main logical operator is the general quantifier. Clearly, the informal proof for one or the other statement may involve an appeal to the elimination rules for $\Rightarrow$ and $\forall$, respectively. But, if that were so, we would have an automatic means of obtaining a proof of the consequent of the conditional or the instantiation of $\forall_x A(x)$. More precisely, whatever proof we accept as an informal proof of $A \Rightarrow B$, merely appending to its last line the rule of $\Rightarrow$-elimination rule yields a proof of B. Similarly, given that we accept a construction as a correct proof of $\forall_x A(x)$, the application of $\forall$-elimination rule yields a proof of $A(t)$, for any t from the domain. Accordingly, the accepted canons of correctness of informal proofs determine what count as proofs of the conditional and the statement with the general quantifier as the main operator, and that determination is not affected by Heyting's clauses. To deal with this problem, Dummett stipulates that the proofs which Heyting's clauses serve to specify should not appeal in the main deduction to the mentioned elimination rules. A similar reflection makes Dummett ban the elimination rules for $\vee$, $\exists$ and $\neg$ from the main deduction. This is not to say that these rules are incorrect, or that one cannot appeal to them in intuitionistically correct proofs. Rather, the rules stand in need of justification, which moreover is easily produced given that the respective proofs satisfy Heyting's respective clauses. Thus, for instance, if we have a construction that we recognize as transforming each proof of a given statement into a proof of another statement, this justifies the elimination rule for the implication, since clearly we obtain the proof of the consequent by applying the construction to a proof of the antecedent. Now, if we conceive of canonical proofs as proofs in the natural deduction calculus, the stipulation amounts to a requirement that no application of elimination rules may occur in their main deductions, and this means that the formulae occurring in the main deduction increase in complexity as we proceed to the conclusion, none of them exceeding the complexity of the conclusion. Accordingly, the distinction between canonical proof and demonstration is analogous to the distinction between normalized and non-normalized proof of the natural deduction formalization of intuitionistic first order logic. If the normalization property holds, Dummett's second requirement, that is of ranking sentences in accord with their meaning-dependence, is satisfied. However, as Dummett observes, the fact that this property holds good for first order logic does not mean that it should hold in any formalized first order theory of intuitionistic mathematics. (EI, p. 396) Moreover, he agrees that the intuitionist is likely to object to the linking of canonical proofs to normalized proofs in a natural deduction calculus for a mathematical theory, since in the latter case introduction rules for $\forall$, $\Rightarrow$ and $\neg$ take a specific form, whereas a canonical proof of each of these statements is *any* construction, provided that it can be recognized as fulfilling the appropriate condition of Heyting's clauses. (*ibid.*, p. 399)

This brings us to the next problem, namely of conceiving of these operations as being recognizable as transforming the relevant proofs. So far we have assumed that these operations can immediately be recognized as transforming proofs of a certain sort into proofs of a given statement. In general, however, we need to require that a canonical proof of the conditional carries with itself a supplementary proof that shows that the procedure *does* transform the proofs of the antecedent into a proof of the consequent. Thus, we face again the danger of the violation of the anti-holistic postulate since there does not seem to be any limit on the complexity of statements that may occur in such a supplementary proof. The hope that the complexity of canonical proofs can be circumscribed in advance is again shattered.

The remedy that Dummett advocates is recourse to Brouwer's concept of *fully analyzed* proof, that is, a proof whose steps are all broken down into a sequence of steps none of which can be analyzed further, and employing only a limited number of types of inference. (EI, p. 94) The need for such a concept is revealed in attempts to exploit fully the intuitionistic meaning of the implication, the idea being that each statement A should be related to an axiom of the form:

$A \Rightarrow$ there is a fully analyzed proof of A having the form of such-and-such a spread.

These conditionals should be seen as stating what form the proof of the statement may have, depending on its structure. (EI, p. 400) If it were possible to formulate such an axiom for each formula and each sentence occuring in a theorem, their conjunction could be added as a supplementary hypothesis to the theorem. Thus, instead of the proof of statement C, Dummett proposes to consider the proof of the conditional C^*, whose consequent is statement C, and whose antecedent is the conjunction of all the axioms for the formulae that occur in C. Thus, the antecedent will consist of conjuncts like '$A \Rightarrow$ there is a proof of A of such and such a form' or '$\forall_x B(x) \Rightarrow$ there is a proof of $B(x)$ of such and such a form', where A and $B(x)$ are constituents of C. According to this proposal, as Dummett argues, one may regard the canonical proof of C as analogous to normalized proofs of the conditional C^*, the consequence being that a bound on the complexity of canonical proofs is restored. (EI, p. 400)

As Dummett concedes, however, the idea of such a linking of statements of intuitionistic mathematics with axioms delineating what their proofs may be is purely programmatic. It is not even known how to arrive at such axioms for intuitionistic arithmetic. Moreover, as he also observes, the success of this program will likely involve the achievement of a complete axiomatization of intuitionistic mathematics, which is a highly spurious outcome due to both the Gödel incompleteness results and a belief that the intuitionist are in particular prone to, namely, that possible methods of a branch of mathematics cannot be circumscribed in advance by any formalization. (*ibid.*, p. 401) All this means that at present there is no reason to think that a meaning theory for the language of mathematical discourse that is framed in terms of intuitionistic provability is any better in respect to anti-holistic features than one appealing to classical provability. Quite apart from this conclusion, one may question the idea that the language of a mathematical discourse is non-holistic. Perhaps this requirement is realistic when applied to natural languages. We may then say that knowledge of language is gradable, adding that someone's failure to understand a lengthy and intricate sentence does not testify to that person's not understanding a simple sentence that shares some expressions with the complex one. Contrary to that, it may be argued that if only a language of a mathematical discourse is sufficiently precise, its understanding is an all-or-nothing matter. A person's inability to use a certain construction, or even worse, to recognize a symbol as belonging to that language, strongly counts against his competence in using the language of a given mathematical theory. Thus, much remains to be done to defend Dummett's anti-holistic postulate, and even more, to fulfill the mathematical program that may provide grounds for preferring a meaning theory framed in terms of the intuitionistic proof to one framed in terms of proof of classical mathematics.

5. RESUME OF DUMMETT'S ARGUMENT

Dummett's case for intuitionism emerges from his program of indirect justification of semantic theories by means of evaluating the meaning theories erected on their basis. On the assumption of this program he argues that the semantic theory that justifies the forms of reasoning applied in classical mathematics does not provide a basis for an acceptable meaning theory. The adjective 'acceptable' points to the postulate that meaning theory should allow for linguistic knowledge to be ascribed to competent language users, in particular, that meanings should be communicable.[51] Given this, the argument can be opposed on three counts. First, one may object to Dummett's *program* rejecting some spurious principles inherent to it or issuing doubts in respect to conclusions reached on such a highly meta-theoretical level. Secondly, one may argue that truth-conditional meaning theory does not fare so badly as Dummett makes out. Finally, one may point out that purported advantages of a meaning theory framed in terms of intuitionistic provability over the truth-conditional account are illusory.

To start with the objections of the first sort, a logically minded person may already scorn the first move of the reasoning, namely the idea that a semantic theory stands in need of some extra-logical justification, his contention being that a semantic theory serves only the tasks of showing, if it is possible, that a given formalization of logic is complete and sound in respect to it, whereas links between the theory and an account of meaning are rather tenuous. Similarly, one may object to the identification of meaning theory with a model of understanding that should allow one to ascribe, in a reasonable way, a grasp of meaning to individuals. Finally, having agreed that a semantic theory should be subjected to justification in accord with whether or not it may serve as the basis for a meaning theory, it may be noticed that the intuitionist does not have anything comparable to classical semantics. Although Heyting's clauses can be seen as specifying the meanings of the logical connectives, as they stand they do not in themselves make up a semantic theory of the sort needed for proofs of the completeness and soundness of intuitionistic logic. Such proofs are given in respect to the semantics of Beth trees, though, but this in turn does not satisfy a principle that links senses to semantic values.

Turning to the objections of the second sort, the first problem is to justify Dummett's narrow concept of *full manifestability*. Indeed, if it is only noted that manifestability of someone's knowledge is in most cases conditional on circumstances, his contention that knowledge of truth conditions is not fully manifestable turns out invalid. Further, while trying to discover what rationale one might have for such a narrow construal of this concept, we have argued that it is not provided by anti-holistic postulates, but rather by a methodological principle that he believes meaning theories should satisfy; however, there is no ground to uphold this principle. Moreover, this construal of manifestability leads to absurd consequences in the case of a currently undecided statement that is proved (or disproved) some time later. The reasoning fails also to show that truth-conditional meaning theory presupposes a hidden ingredient of meaning that transcends use. Accordingly, it cannot be said that the truth-conditional theory violates the demand of intersubjectivity of meaning.

To come finally to the alleged advantages of a meaning theory framed in terms of intuitionistic proof, it turns out that they are far from being established, to say the least. Both theories will face the rather sticky problem of the ascription of knowledge, if a statement in question remains undecided in the relevant period. More importantly, in order to meet anti-holistic postulates, we need to arrive at a viable concept of canonical proof. However, Dummett ideas of linking canonical proofs either to normalized proofs in the natural deduction formalization of a given mathematical theory or to Brouwerian fully analyzed proofs are merely programmatic. Moreover, one may speculate that if realized, they would conflict with Gödel incompleteness results.

To conclude, Dummett's case for intuitionism fares rather badly on both counts: It is unconvincing in arguing that the truth conditional meaning theory makes meaning incommunicable. It is even worse in its attempt to show why the intuitionistic meaning theory should be any better than its classical rival.

5
CONCLUSIONS

In this chapter we shall take stock of the arguments investigated and answer the questions posed in the introduction. Concentrating upon the issue of the intersubjectivity of mathematical knowledge, we have dealt with the three perhaps best known positions that argue for the intuitionistic revision of mathematics: Brouwer's, Heyting's, and Dummett's. While investigating Brouwer's and Heyting's conceptions of intuitionistic mathematics, our query was whether the way they conceive of mathematical constructions, the language of mathematical discourse or the status of laws of logic leads to the consequence that mathematical results are incommunicable. This query has a purely philosophical character, as it concerns conceptions of mathematics and their possible philosophical consequences, neglecting altogether questions of how well intuitionists communicate and whether they do so any better than their rivals. Now, there are two tenets of Brouwer's and Heyting's philosophy that purportedly rule out the communicability of mathematical results. The first stresses the mental character of mathematical constructions, arguing that these should be created by means of the intuition, rather then introduced in a language-dependent way. The second tenet adds that the exactness of mathematics cannot be secured by linguistic means, seeing that language is an irreparably imperfect medium for both the communication and the description of mathematical constructions. Now, there is a certain irony in the fact that the question of intersubjectivity also enters the scene in the opposing argument that Dummett makes against classical logic and mathematics. More precisely, his reasoning purports to show that incommunicability of meaning follows from a meaning theory based on classical semantics. In turn, as his program assumes the need to justify semantic theories by the meaning theories erected on them, he calls for abandoning classical semantics. To get a grip on what is involved in the mutual allegations, we have turned to a reflection on what it is to say that something is intersubjective, ending by specifying two conditions for intersubjectivity, the mentalist and the Wittgensteinian. The former requires that, in order to communicate people's perceptual contents, thoughts or memories should contain some invariant ingredients common to all thinking human subjects, whereas the latter demands that in the process of language-learning people should be able to acquire abilities to use expressions in basically the same way, their agreement in the use of an expression being agreement in their grasp of its meaning. It appears that the mental character of mathematical constructions taken together with the assumption that they serve as denotations of mathematical expressions is at variance with the latter condition. Another view that seems to conflict with that condition consists in assuming that there is something more to the meaning of an expression, for instance some mental content, that cannot be fully manifested in the use of the expression.

As to the first condition we argued that both Heyting's and Brouwer's conceptions of mathematics satisfy it, although the former does so in a somewhat *ad hoc* manner. Brouwer draws a distinction between the mathematical relation and its subjective representation, the latter being unaffected by idiosyncrasies of the individual mathematician, as it is introduced by the intuition of two-ity that is assumed to be identical for all thinking subjects. Heyting, on the other hand, points to the alleged psychological fact that our thoughts concerning (small) natural numbers are 'analogous'. For both conceptions the greater problem is meeting the second condition. Nevertheless, we have argued that mere adherence to the mental character of mathematical constructions does not result in the violation of this condition, the true culprit being rather the view that the constructions must be somehow privately named. There is no reason, however, to hold that Brouwer or Heyting succumbed to this view. This view imposes itself only if it is believed that mathematical expressions *refer*, but again we have argued that the intuitionist is not forced to hold to this view. Besides, the assumption of the referential character of mathematical expressions is equally troublesome for the Platonist, since without postulating that people 'intuit' mathematical objects in basically the same way, the Platonist faces a similar enigma of why we communicate about mathematical matters. Turning now to the opposite argument, namely Dummett's case against classical semantics, it only hints at a certain conundrum arising from the demand that the meaning theory should permit us reasonably to ascribe to people the grasp of the truth conditions of yet-undecided statements, where the assumption is that knowledge of the truth condition should be fully manifestable. The argument fails, however, to show that the proponent of the truth-conditional account is committed to a belief in a hidden ingredient of meaning. Moreover, the conundrum disappears once it is noticed that the manifestability of knowledge is conditional—the grasp of truth conditions can be manifested but only if appropriate circumstances obtain.

In our second question we asked whether the intuitionists have a meaning theory that could be seen as offering grounds for rejecting the concept of bivalent truth. With Brouwer's occasional remarks on meaning of mathematical statements being too scarce to attribute to him such a strategy, the first intuitionistic account of meaning, drawing on Husserl's and Becker's teaching, is to be found in Heyting's writings. However, in order for this account to tell in favor of the intuitionistic meanings of the logical constants, its supplementary premises concerning choice sequence, the potential character of infinity, and freedom of generation of such sequences must be established. Thus, to argue for the correctness of this semantical account, one needs first to have convincing grounds for repudiating bivalent truth in reasoning pertaining to choice sequences, which means that it is not Heyting's semantic theory that leads to a revision of the concept of truth, but the other way round. Finally, to comment on Dummett's case, it is some principles that he believes meaning theories should satisfy which that he sees as telling against bivalent truth.

Let us finally turn to the most galvanizing question, namely the assessment of the arguments for intuitionism. To reflect on the situation, there is well entrenched classical mathematics, with the forms of reasoning of classical logic, the use of actual infinity and a number of other notions elaborated in the history of mathematics. An attempt at revising this subject is made by a small group of mathematicians who claim its methods and canons of reasoning are faulty. The fault of classical mathematics, however, is not seen in any purported inconsistency of classical mathematics or in paradoxes or antinomies that it allegedly generates. The investigated cases for intuitionism accuse the classical mathematician of unjustified reliance on laws of logic (Brouwer), improperly conceiving of the nature of infinity (Brouwer, Heyting), reliance in mathematical reasoning on Platonist metaphysical assumptions (Heyting), or using a semantic theory that, on the assumption of a rather special philosophical program, endangers the intersubjectivity of meaning (Dummett). This character of the arguments is its principal weakness as they may be of some appeal only to a fundamentalist, that is, to a person who may be ready to concede that there is no inconsistency in classical mathematics, that it is powerful and even beautiful, but holds nevertheless that it misrepresents the truth. Even worse, as we have shown, the arguments investigated fare rather badly, either because there is a mistake in an essential step or because they draw on doctrines which are not only alien but also hardly convincing to classical mathematicians. Thus, in Dummett's argument, even if we set aside his controversial program of the justification of semantic theories, the essential claim that the truth-conditional account makes meaning incommunicable is mistaken. In turn, Brouwer's criticism of classical mathematics according to which it relies on language-based ways of introducing mathematical objects has little persuasive power, due to both the vagueness of the distinction between language-based and intuitive, and a rather unusual view of logic. Perhaps most promising is the argument, which we attributed to both Brouwer and Heyting, that tries to show that in reasoning pertaining to choice sequence we need to understand the logical constants in the intuitionistic way, it being further assumed that choice sequence is a general notion for infinite sequences, its special case being sequences determined by laws of progression. It is, however, plain in this case as well, since choice sequences have no place in classical mathematics, that there is much more to be done to make the argument convincing.

APPENDIX

We show that a contradiction follows from the assumption that for any choice sequence of natural numbers it can be decided whether or not it consists of zeros only; in symbols

$$\neg \forall_\alpha (\forall_k \alpha(k) = 0 \lor \neg \forall_k \alpha(k) = 0),$$

where Greek letters $\alpha, \beta, \gamma, \ldots$ range over choice sequences of natural numbers, $k = 0, 1, 2, 3, \ldots$, symbol $\alpha(n)$ stands for the n-th element of α and α_n denotes the sequence $\alpha(0), \alpha(1), \alpha(2), \ldots, \alpha(n-1)$. The presentation follows Dummett's argument, as it occurs in his (1977) book.

1. THE MEANING OF $\forall\exists$

To begin, let us consider how the intuitionist understands the assertion of $\forall_\alpha \exists_n B(\alpha, n)$. Intuitionistically it can only mean that, given any choice sequence α, a natural number n can be found such that the relation $B(\alpha, n)$ holds. If this is to be possible, however, in calculating the number n one needs to take into account only some initial segment of α together with all restrictions on the further elements of α that are imposed during the process of generating this segment. Thus, $\forall_\alpha \exists_n B(\alpha, n)$ asserts the possibility of finding an intensional function ψ, which on the grounds of some initial segment of α and the imposed restrictions produces a number $n = \psi(\alpha)$, such that $B(\alpha, \psi(\alpha))$ holds. In short, $\forall_\alpha \exists_n B(\alpha, n) \Rightarrow \exists_\psi \forall_\alpha B(\alpha, \psi(\alpha))$. By 'intensional' it is meant here that the calculation of n for a sequence α is based on (1) an initial sequence of elements of α, and (2) restrictions introduced in the process of generating the initial sequence and possibly (3) the order of introducing the restrictions. Our first aim is to show that this intensional function can be represented by a function from *infinite* sequences of natural numbers to natural numbers. We will do so in sections 4 and 5 below.

2. EXTENSIONALITY

In intuitionism one encounters both extensional identity and intentional identity. To focus on choice sequences, it is said that two choice sequences are extensionally identical if and only if they have the same elements,

$$\alpha =_{\text{ext}} \beta \text{ if and only if } \forall_k\, \alpha(k) = \beta(k),$$

whereas intensional identity requires both the choice sequence to be generated in the same way. We further say that the relation $B(\alpha, n)$ is extensional in respect to α, in short $\text{Ext}_\alpha B(\alpha, n)$, meaning that if we know α to be extensionally identical to β and we know that $B(\alpha, n)$, then we can show that $B(\beta, n)$. Clearly, if $B(\alpha, n)$ is extensional in respect to α for a given n, then it is extensional in respect to α for any n. Now, one may view the process of generating a choice sequence as consisting in introducing, at any stage, a number together with a restriction imposed on the generation of subsequent numbers. Thus, the choice sequence is represented as the sequence of pairs:

$$\langle \alpha(0), R_0^\alpha \rangle, \langle \alpha(1), R_1^\alpha \rangle, \langle \alpha(2), R_2^\alpha \rangle, \langle \alpha(3), R_3^\alpha \rangle, \ldots$$

where R_k^α is the restriction imposed at the k-th stage of the generation. If one knows that α is extensionally identical to β, he must come to this knowledge by examining only some initial sequences of pairs

$$\langle \alpha(0), R_0^\alpha \rangle, \ldots \langle \alpha(k), R_k^\alpha \rangle \text{ and } \langle \beta(0), R_0^\beta \rangle, \ldots, \langle \beta(m), R_m^\beta \rangle.$$

Without loss of generality we may, however, adopt the convention that restrictions 'carry over' from the stage at which they were introduced to any later stage. For instance, if at stage k the requirement is accepted that any number generated from the stage k onward must be smaller than 10, and later at stage l another requirement demands the generated elements to be even, we say that the restriction R_l^α requires any element from l-th onwards to be even *and* smaller than 10. Thus, with this convention any restriction comprises all earlier restrictions. Granted this convention, a proof that two choice sequences are extentionally identical or a proof that a choice sequence has an extensional property does not draw on the order of imposed restrictions. This leads to the following principle of data.

3. THE PRINCIPLE OF DATA

Assume that a sequence

$$\langle \alpha(0), R_0^\alpha \rangle, \langle \alpha(1), R_1^\alpha \rangle, \langle \alpha(2), R_2^\alpha \rangle, \langle \alpha(3), R_3^\alpha \rangle, \ldots$$

is being generated. Suppose that from its initial part

$$\langle \alpha(0), R_0^\alpha \rangle, \langle \alpha(1), R_1^\alpha \rangle, \langle \alpha(2), R_2^\alpha \rangle, \ldots, \langle \alpha(k), R_k^\alpha \rangle$$

we know that an extensional property $A(\alpha)$ holds. This knowledge must derive from investigating the initial sequence α_k and the 'final' restriction R_k^α. Consequently, if R_k^α imposes a requirement on the future generation of α, say $C(α)$, then for any choice sequence γ that shares with α the initial segment $α(k)$ and satisfies the requirement $C(γ)$, the property $A(γ)$ holds. Symbolically,

$$\forall_{γ \in α_k} (C(γ) \Rightarrow A(γ)),$$

where the notation $γ \in α_k$ indicates that γ has the initial segment $α_k$.

4. REPRESENTING

We shall show that the intentional function ψ may be represented by a function e from finite sequences of natural numbers to natural numbers. Following Dummett's notation (1977) we say that

DEFINITION 1
e represents ψ iff for any α one can find a number n such that

$$e(α_n) = ψ(α) + 1 \text{ and for all } m < n \ e(α_m) = 0,$$

where we leave unspecified the value $e(α_k)$ for $k > n$. The equation $e(α_m) = 0$ indicates that the initial segment $α_m$ is too short to calculate $ψ(α)$. The first finite segment of α for which e gives a positive number is sufficient to determine $ψ(α)$. It is convenient to assume the following definition:

DEFINITION 2

$$e^*(α) \text{ is defined iff } \exists_n e(α_n) > 0$$
$$e^*(α) = k \text{ iff } \exists_n (e(α_n) = k+1 \wedge \forall_{m<n} e(α_m) = 0)$$

With this definition, if $e^*(α)$ is defined, then $e^*(α) = ψ(α)$.

5. ∀∃-CONTINUITY PRINCIPLE

The argument for the continuity principle invokes the concept of lawless choice sequence. The generation of such sequences proceeds freely as no restrictions are imposed. Now, assume that $B(α, n)$ is extensional and that $\forall_α B(α, ψ(α))$ holds. Let β be a lawless choice sequence. If it is lawless, than the calculation of $ψ(β)$ must be based on an initial segment $β_k$ only, as the generation is not subject to any restriction. Thus, at least for all lawless β the intensional function ψ can be represented by a function e^* such that $ψ(β) = e^*(β)$. We will argue moreover that if e^* is defined for lawless choice sequences, then the continuity principle holds.

Suppose thus $\psi(\beta) = e^*(\beta)$ for any lawless β. Assume further that $e(\alpha_k) = n+1$. Since e^* is defined for lawless sequences and given definition 2, if β is lawless and $\beta \in \alpha_k$, then $B(\beta, n)$ holds. However, to arrive at the knowledge that $B(\beta, n)$ holds, we need to take into account both the segment α_k and the set of restrictions imposed during its generation. But to say that β is lawless means that no restriction is imposed on its generation. Thus, applying the principle of data, we get

$$\forall_{\gamma \in \alpha_k} B(\gamma, n),$$

which, given $e(\alpha_k) = n+1$ and definition 2, amounts to

$$\forall_{\gamma \in \alpha_k} B(\gamma, e^*(\alpha)). \tag{1}$$

From Eq. 1 we deduce the $\forall\exists$-continuity principle:

$$\forall_n \text{Ext}_\alpha B(\alpha, n) \wedge \forall_\alpha \exists_n B(\alpha, n) \Rightarrow \exists_{e^*} \forall_\alpha [e^*(\alpha) \text{ is defined} \wedge B(\alpha, e^*(\alpha))]. \tag{2}$$

This ends the argument for the $\forall\exists$-continuity principle. The reasoning presented here follows Dummett's argumentation. (1977, pp. 447–448) Two of its features are perhaps worth pointing out. First, it is crucial for the argument to assume that lawless sequences are intuitionistically legitimate objects. Secondly, it relies on a method rather peculiar to intuitionism, namely a reflection on how one could know that a premise holds.

6. CONTINUITY PRINCIPLE FOR DISJUNCTION

From the $\forall\exists$-continuity principle just established we prove its version for disjunction. Let us suppose that $\forall_\alpha(C(\alpha) \vee D(\alpha))$ holds, which means that for any α we can find out whether $C(\alpha)$ or $D(\alpha)$ holds. Thus, we may introduce a number n such that if $D(\alpha)$ does not hold, then $n = 0$, but if $D(\alpha)$ holds, then $n = 1$. Since if $D(\alpha)$ does not hold, $C(\alpha)$ must hold, we obtain:

$$\forall_\alpha \exists_n ((C(\alpha) \wedge n = 0) \vee (D(\alpha) \wedge n = 1)).$$

Given that the properties $C(\alpha)$ and $D(\alpha)$ are extensional, the relation

$$(C(\alpha) \wedge n = 0) \vee (D(\alpha) \wedge n = 1)$$

is also extensional. For, if we have a proof of $C(\alpha)$ or a proof of $D(\alpha)$ and moreover a proof of $\alpha =_{\text{ext}} \beta$, we can transform them into a proof of

$$(C(\beta) \wedge n = 0) \vee (D(\beta) \wedge n = 1).$$

This means that the following implication holds:

$$\text{Ext}_\alpha C(\alpha) \wedge \text{Ext}_\alpha D(\alpha) \wedge \forall_\alpha (C(\alpha) \vee D(\alpha))$$
$$\Rightarrow \forall_n \text{Ext}_\alpha ((C(\alpha) \wedge n = 0) \vee (D(\alpha) \wedge n = 1))$$
$$\wedge \forall_\alpha \exists_n ((C(\alpha) \wedge n = 0) \vee (D(\alpha) \wedge n = 1)). \quad (3)$$

Accordingly, plugging the consequent of Eq. 3 into the premise of $\forall \exists$-continuity principle (Eq. 2) we obtain the continuity principle for disjunction:

$$\text{Ext}_\alpha C(\alpha) \wedge \text{Ext}_\alpha D(\alpha) \wedge \forall_\alpha (C(\alpha) \vee D(\alpha)) \Rightarrow$$
$$\exists_{e^*} \forall_\alpha [e^*(\alpha) \text{ is defined } \wedge ((C(\alpha) \wedge e^*(\alpha) = 0) \vee (D(\alpha) \wedge e^*(\alpha) = 1)]. \quad (4)$$

7. CONTRADICTORITY OF AN INSTANCE OF THE GENERALIZED EXCLUDED MIDDLE

In the final step let us put for $C(\alpha)$: $\forall_k \alpha(k) = 0$ and for $D(\alpha)$: $\neg \forall_k \alpha(k) = 0$ in Eq. 4. Both being extensional, we obtain:

$$\forall_\alpha (\forall_k \alpha(k) = 0 \vee \neg \forall_k \alpha(k) = 0) \Rightarrow \exists_{e^*} \forall_\alpha [e^*(\alpha) \text{ is defined}$$
$$\wedge ((\forall_k \alpha(k) = 0 \wedge e^*(\alpha) = 0) \vee (\neg \forall_k \alpha(k) = 0 \wedge e^*(\alpha) = 1))]. \quad (5)$$

The consequent of this implication is absurd, however. Let us assume first that for a certain β, $\beta(k) = 0$ for any k and $e^*(\beta) = 0$. Given definition 2, we may thus find an n such that

$$e(\beta_n) = 1 \text{ and } \forall_{m<n} e(\beta_m) = 0.$$

But this means, since e^* is defined for any choice sequence, that any choice sequence α sharing with β its initial sequence β_n consists of zeros only (because by definition 2, $\forall_{\gamma \in \beta_n} e^*(\gamma) = 0$). This is absurd, however, as we may construct the following choice sequence α:

$$\alpha(k) = \beta(k) = 0 \quad \text{if } k < n,$$
$$\alpha(k) = 1 \quad \text{if } k \geq n.$$

Thus, no choice sequence can satisfy the first disjunct in the consequent of Eq. 5.
Consequently, for any α

$$\neg \forall_k \alpha(k) = 0 \wedge e^*(\alpha) = 1.$$

This implies that no choice sequence can consist of zeros only. But this is again absurd, since clearly we can construct a sequence consisting of zeros only. We have thus found that a contradiction follows from assuming the consequent of implication from Eq. 5. Thus, the antecedent of this implication cannot hold either, so we have proved that

$$\neg \forall_\alpha (\forall_k \alpha(k) = 0 \vee \neg \forall_k \alpha(k) = 0).$$

NOTES

[1] Compare for instance Dummett PBIL, p. 226 or Tait (1983), pp. 176–177.

[2] The most recent and comprehensive exposition of Martin-Löf's theory is his (1996) paper; some other works of his that are worth consulting are listed in the bibliography.

[3] I owe this point to Paweł Turnau.

[4] For any reader who is interested in a more historical perspective on both Brouwer's philosophy and mathematics, van Stigt's (1990) detailed work is highly recommended.

[5] As documented in VS, Brouwer mentions Schopenhauer in his letters to Korteweg. Also Brouwer's concept of will (see his 1929 and 1933 papers) has strikingly Schopenhauerian overtones.

[6] See his (1948) paper.

[7] Brouwer's term appears at first sight cumbersome, as it is natural to think that he had in mind simply the movement of time, or its flow. Nevertheless, one may suspect that his choice of terminology was deliberate, as he may have wanted to stress the discrete character of the motion. Indeed, his basic concept of one sensation giving way to another has a discrete character. The continuum emerges once the attention is focused on what is between the mentioned sensations; Brouwer calls it the 'intuitive between'.

[8] Van Dalen (1978, p. 302) points to two somehow anecdotal facts from Brouwer's life. Despite his critical appraisal of the natural sciences and his belief that the laws of nature are human constructs, he lectured in mechanics, of all things, for several years. Moreover, during World War I he prepared and later published two cartographic works which he believed could help the Dutch army.

[9] Brouwer uses also the term 'basic intuition of two-ity' or 'intuition of *two-oneness*'.

[10] English translation in Kant (1958).

[11] For a discussion of Brouwer's debts to Kant, see also Posy (1974) and (1984).

[12] Compare (1952), p. 511.

[13] See Brouwer (1981), p. xi, van Dalen's preface.

[14] I owe these points to Paweł Turnau.

[15] A reader interested in the origin of the notion of choice sequence is referred to Troelstra (1982). For getting acquainted with mathematics of choice sequences, Troelstra (1977) and (1983) are recommended.

[16] It is described in van Dalen's commentary in Brouwer (1981), p. 17.

[17] For details of such an enumeration, see Brouwer (1954), p. 529.

[18] A reader interested in both the development and details of Brouwer's concept of species is referred to van Stigt's excellent study (VS), pp. 335–356.

[19] See Cantor (1932), pp. 139–266.

[20] Compare *ibid.*, p. 197.

[21] Cantor distinguishes between two ways mathematical objects are real. Their *intrasubjective* reality consists in their occurring in thinking of human subjects, whereas their *transubjective* or *transient* reality is supposedly related to their Platonistic status.

22 Compare Bunge (1962), p. 43.

23 In his 'Mathematische Probleme'. See Hilbert (1901), p. 297.

24 A more elaborate argument to the effect that the notion of truly undecidable statement is not meaningful for the intuitionist can be found in Martin-Löf (1995).

25 Brouwer's involvement in significs is described in van Stigt (1982).

26 This is how I translate Brouwer's *mathematische Betrachtung* from his (1929) paper.

27 Hilbert did not give a precise definition of what real and ideal sentences are; nor was the finitary vs. infinitary distinction very precise. For this reason his program allows for several interpretations—see Murawski (1994).

28 *The Thirteen Books of Euclid's Elements.*

29 Quoted from Heyting's interview included in VS, p. 286.

30 In the syllogism, we should rather have predicates (general terms) than individual terms like 'Socrates'. Thus, Brouwer's example should be corrected in the following way: If all men are mortal and all Greeks are men, then all Greeks are mortal.

31 One may wonder how to square this view of Brouwer's with his much later, rather positive assessment of intuitionistic logic (1955) and the fact that Arend Heyting, his student and then close collaborator, developed intuitionistic logic under his supervision. This subject is treated at length in van Stigt (1990) book, and for this reason we do not enter this discussion here.

32 Included in VS, pp. 502–505.

33 Compare his (1929), p. 424 and also (1933), in VS p. 426.

34 According to van Stigt (VS, p. 204), the original Dutch version simply states: "Expected experiences as such and the reputed experiences of others as such are not truths (...) There are no non-experienced truths."

35 Compare Brouwer's manuscript 'Changes in the Relation Between Classical Logic and Mathematics', as translated and published by van Stigt, VS, pp. 453–58.

36 It was first delivered as the Rector's address at the opening ceremony of the 1922 academic year at Warsaw University and later, in 1946, prepared for publication; the English translation appeared in Łukasiewicz (1970).

37 For a recent discussion of psychologism, see Kutsch (1995).

38 Brouwer answers to Griss's project in (1948a) pointing to a negative property that cannot be translated into a positive one.

39 See Heyting (1934), p. 388.

40 Griss's objections and Heyting's replies to them are discussed in Troelstra (1983b).

41 Compare Heyting (1956), p. 17.

42 *Through the Looking-Glass*, chapter VI.

43 Compare Heyting (1959), p. 70.

44 Still, it is controversial whether Heyting could agree on this view. One may read the discussed quotation as merely a perception that Brouwer's total neglect of mathematical language should be somehow amended, without conceding any autonomous role to the rules of the language.

45 The other option is to explain the validity in terms of ordering, which is typical for semantic theories operating with relativised truth values (see section 2.2). A theory of this sort assumes a space equipped with an ordering such that points of this space are assigned to formulae and then states that a form of argument is valid if under any assignment the appropriate order relations hold (LBM, p. 43).

46 Compare LBM, pp. 33–35.

47 See EI, pp. 190–193 and van Dalen (1986), p. 247–250.

48 A case may be put for relaxing the requirements for the adequacy of meaning theory, replacing Dummett's postulate that the theory should contain a part explaining how non-

statable knowledge of meaning should be manifested by weaker demands—see for instance McDowell (1977) or Prawitz (1978).

[49] For our reasons for adopting this terminology, see chapter 2 section 2.4.

[50] There are, however, arguments against the identification of the meaning of a sentence, understood as the grasp of its truth conditions, with the possible use of the sentence. See for instance McGinn (1980) and (1982).

[51] Apart from this argument, sometimes called 'semantical', there is also another one ('proof theoretical') that has much to do with Prawitz's works on justification of the rules for the logical constants and Dummett's emphasis on the harmony that should obtain between aspects of the use of sentences. We mention it only briefly as it lies rather on the periphery of the present considerations. For a discussion of harmony requirements see Dummett's EI and LBM; Prawitz's justification procedures are discussed in his (1973) and (1977).

BIBLIOGRAPHY

Appiach, A.: 1985, 'Verificationism and the Manifestation of Meaning', *Aristotelian Society Supplementary Volume* **59**, pp. 17–52.
Banaceraff, P., Putnam, H. (eds.): 1983, *Philosophy of Mathematics: Selected Readings*, 2nd edition, Cambridge University Press, Cambridge.
Becker, O.: 1927, *Mathematische Existenz*, in *Jahrbuch für Philosophie und Phönomenologische Forschung*, vol. VIII.
Beth, E.W.: 1965, *The Foundations of Mathematics*, North Holland, Amsterdam.
Brouwer, L.E.J.: 1975, *L.E.J Brouwer Collected Works*, vol. 1, ed. by A. Heyting, North Holland, Amsterdam, referred to as *BCW*.
Brouwer, L.E.J.: 1905, 'Leven, Kunst, Mistiek' (Life, Art, Misticism), in *BCW*, pp. 1–10.
Brouwer, L.E.J.: 1907, 'On the Foundation of Mathematics', in *BCW*, pp. 11–101.
Brouwer, L.E.J.: 1908, 'The Unreliability of the Logical Principles', in *BCW*, pp. 107–111.
Brouwer, L.E.J.: 1919, 'Signifische Sprachforschung', in *BCW*, pp. 222–229.
Brouwer, L.E.J.: 1929, 'Mathematik, Wissenschaft und Sprache', in *BCW*, pp. 417–428.
Brouwer, L.E.J.: 1933, 'Volition, Knowledge, Language', in van Stigt (1990), pp. 419–431.
Brouwer, L.E.J.: 1937, 'Signific Dialogues', in *BCW*, pp. 447–542.
Brouwer, L.E.J.: 1946, 'Synopsis of the Signific Movement in the Netherlands. Prospects of the Signific Movement', in *BCW*, pp. 465–471.
Brouwer, L.E.J.: 1947, 'Guidelines of Intuitionist Mathematics', in *BCW*, p. 477.
Brouwer, L.E.J.: 1948, 'Consciousness, Philosophy and Mathematic', in *BCW*, pp. 480–494.
Brouwer, L.E.J.: 1948a, 'Essentially Negative Properties', in *BCW*, pp. 478–479.
Brouwer, L.E.J.: 1952, 'Historical Background, Principles and Methods of Intuitionism', in *BCW*, pp. 508–515.
Brouwer, L.E.J.: 1954, 'Points and Spaces', in *BCW*, pp. 522–538.
Brouwer, L.E.J.: 1955, 'The Effect of Intuitionism on Classical Algebra of Logic', in *BCW*, pp. 551–554.
Brouwer, L.E.J.: 1979, 'The Rejected Parts of Brouwer's Dissertation on the Foundations of Mathematics', ed. by P. van Stigt *Historia Mathematica* **6**, pp. 385–404.
Brouwer, L.E.J.: 1981, *Brouwer's Cambridge Lectures on Intuitionism*, ed. by D. van Dalen, Cambridge University Press, Cambridge.
Bunge, M.: 1962, *Intuition and Science*, Prentice–Hall, Englewood Cliffs, N.J.
Burgess, J.: 1984, 'Dummett's Case for Intuitionism', *History and Philosophy of Logic* **5**, pp. 177–194.
Cantor, G.: 1932, *Gesammelte Abhandlungen mathematischen und philosophischen Inhalts*, Springer Verlag, Berlin.
Caroll, L.: 1994, *Through the Looking-Glass*, Penguin Books, London.
Clark, P.: 1993, 'Logicism, the continuum and anti-realism', *Analysis* **53** (3), pp. 12–141.
Dąmbska, I.: 1976, 'Idee Kantowskie w neointuicjonizmie Brouwera' (Kantian themes in Brouwer's neointuitionism), *Acta Universitatia Wratislaviensis* **290**, pp. 7–14.
Dąmbska, I.: 1976a, 'Kanta filozofia matematyki i kontynuacja niekórych jej myśli w twórczości H. Poincarego' (Kant's philosophy of mathematics and the continuation of some of

its strands in H. Poincare's thought), in J. Garewicz (ed.): *Dziedzictwo Kanta* (The Legacy of Kant), PWN, Warsaw, pp. 287–307.
Davidson, D.: 1967, 'Truth and Meaning', *Synthese* **7**, pp. 304–323.
Devitt, M.: 1983, 'Dummett's Anti-Realism', *Journal of Philosophy* **80**, pp. 73–99.
Dummett, M.: 1977, *Elements of Intuitionism*, Oxford University Press, Oxford, referred to as *EI*.
Dummett, M.: 1978, *Truth and Other Enigmas*, Duckworth, London, referred to as *TOE*.
Dummett, M.: 1975, 'The Philosophical Basis of Intuitionistic Logic', in *TOE*, pp. 215–247, referred to as *PBIL*.
Dummett, M.: 1991, *The Logical Basis of Metaphysics*, Harvard University Press, Cambridge Mass., referred to as *LBM*.
Dummett, M.: 1976, 'What Is a Theory of Meaning II', in Evans and McDowell (1976), pp. 67–137, referred to as *WTM*.
Dummett, M.: 1963, 'Realism', in *TOE*, pp. 145–165.
Dummett, M.: 1973, 'The Justification of Deduction', in *TOE*, pp. 290–319.
Dummett, M.: 1974, 'The Social Character of Meaning', in *TOE*, pp. 420–430.
Dummett, M.: 1975a, 'What Is a Theory of Meaning I', in Guttenplan (1975), pp. 97–138.
Dummett, M.: 1975b, 'Wang's Paradox', in *TOE*, pp. 248–268.
Dummett, M.: 1975c, 'Can Analytical Philosophy Be Systematic and Ought It to Be', in *TOE*, pp. 437–459.
Dummett, M.: 1975d, 'Frege's Distinction Between Sense and Reference', in *TOE*, pp. 116–145.
Dummett, M.: 1982, 'Realism' *Synthese* **52**, pp. 51–111.
Dummett, M.: 1989, 'More About Thoughts', *Notre Dame Journal of Symbolic Logic* **30**, pp. 1–19.
Dummett, M.: 1993 *Origins of Analytical Philosophy*, Duckworth, London.
Edginton, D.: 1981, 'Meaning, Bivalence and Realism', *Proceedings of the Aristotelian Society* **81**, pp. 153–73.
Edwards, J.: 1995, 'The universal quantifier and Dummett's verificationist theory of sense', *Analysis* **55** (2), pp. 90–97.
Euclid: 1952, *The Thirteen Books of Euclid's Elements*, trans. by Sir T. Heath, Encyclopedia Britannica, Inc., Chicago.
Evans, G., McDowell, J. (eds.): 1976, *Truth and Meaning*, Clarendon Press, Oxford.
Fine, K.: 1975, 'Vagueness, Truth and Logic', *Synthese* **30**, pp. 265–300.
Frege, G.: 1892, 'Über Sinn und Bedeutung', Eng. trans. in *Translations from the Philosophical Writings of Gottlob Frege*, trans. by P. Geach and M. Black, Basil Blackwell, Oxford, 1960.
Frege, G.: 1918, 'Der Gedanke', Eng. trans. in G. Frege: *Collected Papers on Mathematics, Logic and Philosophy*, ed. by B. McGuinness, Basil Blackwell, Oxford 1984, pp. 351–372.
Gabbay, D., Guenthner, F. (eds.): 1986, *Handbook of Philosophical Logic*, vol. III, Reidel, Dordrecht.
Griss, G.F.C.: 1946, 'Negationless intuitionistic mathematics I', *Nederl. Akad. Wetensch. Proc.* **49**, pp. 1127–1133.
Grzegorczyk, A.: 1965, 'Konsekwencje teoriopoznawcze dwóch twierdzeń matematyki' (Epistemological implications of two theorems of mathematics), *Studia Filozoficzne* **3** (42), pp. 115–118.
Grzegorczyk, A.: 1974, *An Outline of Mathematical Logic*, Reidel, Dordrecht.
Guttenplan, S. (ed.): 1975, *Mind and Language*, Clarendon Press, Oxford.
Haack, S.: 1978, *Philosophy of Logics*, Cambridge University Press, Cambridge.
Haldane, J., Wright, C. (eds.): 1993, *Reality, Representation and Projection*, Oxford University Press, Oxford.

Hale, B.: 1997, 'Realism and its opposition' in Wright and Hale (1997), pp. 271–308.
Heyting, A.: 1930, 'Die formalen Regeln der intuitionistischen Logik', *Sitzungsberichte der preussischen Akademie der Wissenschaften, physikalisch–mathematische Klasse*, Berlin, pp. 42–56, 57–71 and 158–169.
Heyting, A.: 1931, 'Die intuitionistische Grundlegung der Mathematik', *Erkentnis* **2**, pp. 106–115, Eng. trans. in Banaceraff and Putnam (1983), pp. 52–61.
Heyting, A.: 1934, *Mathematische Grundlagenforschung. Intuitionismus. Beweistheorie*, Springer, Berlin.
Heyting, A.: 1947, 'Formal Logic and Mathematics', *Synthese* **6**, (1947–48), pp. 275–282.
Heyting, A.: 1956, *Intuitionism. An Introduction*, North Holland, Amsterdam.
Heyting, A.: 1958a , 'Blick von der intuitionistischen Warte', *Dialectica* **12**, pp. 332–345.
Heyting, A.: 1958b, 'Intuitionism in Mathematics', in Klibansky (1958), pp. 101–115.
Heyting, A.: 1959, 'Some Remarks on Intuitionism', in A. Heyting (ed.): *Constructivity in Mathematics*, North Holland, Amsterdam, pp. 67–71.
Heyting, A.: 1962, 'After Thirty Years', in Nagel *et al.*, (1962), pp. 194–197.
Heyting, A.: 1968, 'Intuitionism in Mathematics', in Klibansky (1968), pp. 313–323.
Heyting, A.: 1974, 'Intuitionistic Views on the Nature of Mathematics', *Synthese* **27**, pp. 79–91.
Hilbert, D.: 1901, 'Mathematische Probleme', in his *Gesammelte Abhandlungen* vol. III, Springer Verlag, Berlin, 1935, pp. 290–298.
Hilbert, D.: 1926, 'Über Unendlichkeit', *Mathematische Annalen* **95**, pp. 161–190.
Hintikka, J.: 1973, *Logic, Language–Games and Information*, Oxford University Press, Oxford.
Hintikka, J.: 1978, *Essays on Mathematical and Philosophical Logic*, Reidel, Dordrecht.
Husserl, E.: 1900/1901 *Logische Untersuchungen*, Niemeyer, Halle trans. by J.N. Findlay, Routledge & Kegan Paul, London, 1970, referred to as *LU*.
Kant, I.: 1952, *The Critique of Pure Reason*, trans. by J.M. Meiklejohn, Encyclopedia Britannica, Inc., Chicago.
Kant, I.: 1958, *Selected Pre-Critical Writings and Correspondence with Beck*, trans. by G.B. Kerferd and D.E. Walford, Manchester University Press, Barnes & Noble, New York.
Klaus, G., Buhr, G. (ed.): 1972, *Philosophisches Wörterbuch*, Verlag Enzyklopödie, Leipzig.
Klibansky, R. (ed.): 1958, *Philosophy in the Mid–Century. A Survey*, La Nuova Italia, Florence.
Klibansky, R. (ed.): 1968, *Contemporary Philosophy. A Survey*, La Nuova Italia, Florence.
Kolmogorov, A.N.: 1932, 'Zur Deutung der Intuitionistischen Logik', *Mathematische Zeitschrift* **35**, pp. 58–65.
Kreisel, R.: 1962, 'Foundations of Intuitionist Logic', in Nagel *et al.*, (1962), pp. 198–210.
Kutsch, M.: 1995, *Psychologism*, Routledge, London.
Leśniewski, S.: 1929, 'Grundzüge eines neuen Systems der Grundlagen der Mathematik', *Fundamenta Mathematicae* **XIV**, pp. 1–81.
Łukasiewicz, J.: 1907, 'Logika a psychologia' (Logic and psychology), *Przegląd Filozoficzny* **X**, pp. 489–491.
Łukasiewicz, J.: 1961, 'O determinizmie' (On determinism), in his *Z zagadnień logiki i filozofii* (Themes from Logic and Philosophy), ed. by J. Słupecki, PWN, Warsaw, 1961, pp. 114–126. Eng. transl. in his *Selected Works*, ed. by L. Borkowski, North Holland, Amsterdam, 1970, pp. 110–128.
Martin-Löf, P.: 1987, 'Truth of a proposition, evidence of a judgment, validity of a proof', *Synthese* **73**, pp. 407–420.
Martin-Löf, P.: 1991, 'A path from logic to metaphysics', in *Atti del Congresso Nuovi Problemi della Logica e della Filosofia della Scienza, Viareggio 8–13 gennaio 1990*, vol. II, CLUEB, Bologna, pp. 141–149.

Martin–Löf, P.: 1994, 'Analytic and synthetic judgments in type theory', in P. Parrini (ed.): *Kant and Contemporary Epistemology*, Kluwer Academic Publishers, Dordrecht, pp. 87–99.

Martin–Löf, P.: 1995, 'Verificationism then and now', in W. DePauli–Schimanovich, E. Köhler, and F. Stadler (eds.): *The Foundational Debate: Complexity and Constructivity in Mathematics and Physics*, Kluwer Academic Publishers, Dordrecht, pp. 187–196.

Martin–Löf, P.: 1996, 'On the Meaning of the Logical Constants and the Justifications of the Logical Laws', *Nordic Journal of Philosophical Logic* **1** (1), pp. 3–10.

Martin–Löf, P.: 1996a, 'Truth and knowability: on the principles C and K of Michael Dummett', in H.G. Dales, G. Oliveri (eds.): *Truth in Mathematics*, Clarendon Press, Oxford.

McGinn, C.: 1980, 'Truth and use', in Platts (1980), pp. 19–40.

McGinn, C.: 1982, 'Realist semantics and content ascription', *Synthese* **52**, pp. 113–134.

Murawski, R.: 1986, *Filozofia matematyki* (Philosophy of Mathematics), Wyd. Naukowe UAM, Poznań.

Murawski, R.: 1994, 'Hilbert's Program: Incompleteness Theorems Vs. Partial Realizations', in Woleński (1994) pp. 103–128.

Nagel, E., Suppes, P. and Tarski, A. (eds.): 1962, *Logic, Methodology and Philosophy of Science*, Stanford University Press, Stanford.

Parsons, C.: 1993, 'On Some Difficulties Concerning Intuition and Intuitive Knowledge', *Mind* **102** (406), pp. 233–246.

Peckhaus, V.: 1990, *Hilbertprogramm und Kritische Philosophie*, Vandehoeck & Ruprecht, Göttingen.

Platts, M. de (ed.): 1980, *Reference, Truth and Reality*, Routledge & Kegan Paul, London.

Posy, K.: 1974, 'Brouwer's Constructivism', *Synthese* **27**, pp. 129–159.

Posy, K.: 1984, 'Kant's Mathematical Realism', *Monist* **67**, pp. 115–134.

Prawitz, D.: 1965, *Natural Deduction: A Proof-Theoretical Study*, Almquist & Wiksel Stockholm.

Prawitz, D.: 1970, 'Constructive Semantics', in *Proceedings of the 1st Scandinavian Logic Symposium*, Uppsala, pp. 96–114.

Prawitz, D.: 1973, 'Towards a Foundation of a General Proof Theory', in P. Suppes, L. Henkin, A. Joja and Gr.C. Moisil (eds.) *Logic, Methodology, Philosophy of Science IV*, North Holland, Amsterdam, pp. 225–50.

Prawitz, D.: 1977, 'Meaning and Proofs: on the Conflict between Classical and Intuitionistic Logic', *Theoria* **43**, pp. 2–40.

Prawitz, D.: 1978, 'Proofs and the Meaning and Completeness of the Logical Constants', in Hintikka (1978), pp. 25–40.

Skorupski, J.: 1997, 'Meaning, use, verification' in Wright and Hale (1997), pp. 29–59.

Sundholm, G.: 1983, 'Constructions, Proofs and the Meaning of the Logical Constants', *Journal of Philosophical Logic* **12**, pp. 151–172.

Sundholm, G.: 1986, 'Proof theory and meaning', in Gabbay and Guenthner (1986), pp. 471–506.

Sundholm, G.: 1994, 'Existence, Proof and Truth-Making—A Perspective on the Intuitionistic Conception of Truth', *Topoi* **13**, p. 117–126.

Swart, H. de: 1974, *Intuitionistic logic in an intuitionistic meta-language*, Math. Inst Kath. Univ. Nijmegen.

Tait, W.W.: 1983, 'Against Intuitionism: Constructive Mathematics is Part of Classical Mathematics', *Journal of Philosophical Logic* **12**, pp. 173–195.

Tarski, A.: 1933, *Pojęcie prawdy w językach nauk dedukcyjnych* (The Concept of Truth in the Languages of Deductive Sciences), Warsaw.

Tarski, A.: 1956, *Logic, Semantics, Metamathematics (Papers from 1923 to 1938)*, trans. by J.H. Woodger, Clarendon Press, Oxford.

Teichmann, R.: 1995, 'Truth, Assertion and Warrant', *The Philosophical Quarterly* **45** (179), pp. 7–85.
Tennant, N.: 1987, *Anti-realism and Logic. Truth as Eternal*, Oxford University Press, Oxford.
Tennant, N.: 1995, 'On negation, truth and warranted assertability', *Analysis* **55** (2), pp. 98–104.
Tennant, N.: 1997, *The Taming of the True*, Clarendon Press, Oxford.
Tieszen, R.: 1989, *Mathematical Intuition: Phenomenology and Mathematical Knowledge*, Kluwer Academic Publishers, Dordrecht.
Tragesser, R.S.: 1984, *Husserl and Realism in Logic and Mathematics*, Cambridge University Press, Cambridge.
Troelstra, A.S.: 1977, *The Theory of Choice Sequences*, Clarendon Press, Oxford.
Troelstra, A.S.: 1982, 'The Origin and Development of Brouwer's Concept of Choice Sequence', in Troelstra and van Dalen (1982), pp. 465–486.
Troelstra, A.S.: 1983a, 'Analysing Choice Sequences', *Journal of Philosophical Logic* **12**, pp. 197–260.
Troelstra, A.S.: 1983b, 'Logic in the writings of Brouwer and Heyting', in *Atti del Convegno Internazionale de Storia della Logica*, CLUEB, Bologna.
Troelstra, A.S.: 1988, 'On the early history of intuitionistic logic', in P.P. Petkov: *Proceedings of the Summer School and Conference on Mathematical Logic*, Plenum Press, New York and London, pp. 3–17.
Troelstra, A., van Dalen, D. (eds.): 1982, *The L.E.J.Brouwer's Centenary Symposium*, North Holland, Amsterdam.
Troelstra, A., van Dalen, D. (eds.): 1988, *Constructivism in Mathematics*, North Holland, Amsterdam.
van Dalen, D.: 1978, 'Brouwer: The Genesis of His Intuitionism', *Dialectica* **32**, pp. 291–303.
van Dalen, D.: 1981, *Brouwer's Cambridge Lectures on Intuitionism*, ed. by D. van Dalen, Cambridge University Press, Cambridge.
van Dalen, D.: 1986, 'Intuitionistic Logic', in Gabbay and Guenthner (1986), pp. 225–239.
van Stigt, W.P.: 1979, 'The Rejected Parts of Brouwer's Dissertation on the Foundations of Mathematics', *Historia Mathematica* **6**, pp. 385–404.
van Stigt, W.P.: 1982, 'Brouwer's Signific Interlude', in Troelstra and van Dalen (1982), pp. 505–512.
van Stigt, W.P.: 1990, *Brouwer's Intuitionism*, North Holland, Amsterdam, referred to as *VS*.
Weyl, H.: 1918, *Das Kontinuum. Kritische Untersuchungen über die Grundlagen der Analysis*, Teubner, Berlin/Leipzig.
Wiggins, D.: 1997, 'Meaning and truth-conditions: from Frege's grand design to Davidson's', in Wright and Hale (1997), pp. 1–28.
Wittgenstein, L.: 1958, *Philosophical Investigations*, trans. by G.E.M. Anscombe, Basil Blackwell, Oxford.
Wittgenstein, L.: 1967, *Zettel*, Basil Blackwell, Oxford.
Woleński, J. (ed.): 1994, *Philosophical Logic in Poland*, Kluwer Academic Publishers, Dordrecht.
Wright, C.: 1992, *Truth and Objectivity*, Harvard University Press, Cambridge, Mass.
Wright, C., Hale, B. (eds.): 1997, *Companion to the Philosophy of Language*, Blackwell Publishers, Oxford.

INDEX

a priori, *see* logic, Brouwer's concept of
acts of intuitionism, 29
addition, *see* species, addition of species
algebra
    algebra Z_2, 175
    Boolean algebra, 175, 181
analogous thoughts, 9, 139, 143
ascription of knowledge, 178–180, 184, 185, 193
assertability condition, 45, 48, 73, 76, 78, 79, 122, 136
axiomatic-deductive method
    Brouwer's views on, 54–57, 59

bar, 162, 163
basis for mathematics, 104, 106–108, 111, 144
Becker, 127, 131–135, 144, 195
Beth tree semantics, 153, 154, 156, 157, 161, 162, 164, 193
Bhagavadgita, 17, 18
bivalence, 2, 3, 25, 71, 72, 74–76, 78, 81, 83, 100–102, 118–126, 131, 144, 145, 158–160, 165, 195
Bloemers, 49
Boole, 59
Boolean, *see* algebra
Borel, 49
break, *see* choice sequence
Buhr, 5
Bunge, 204
Burali-Forti paradox, 40

canonical proof, 189–191
    as a fully analyzed proof, 191
    as a normalized proof, 191
Cantor, 34, 36–40, 48

Caroll, 127
causal attention, 19–23, 25–27, 41, 49, 65, 86–88, 100, 103, 106
causal link, 18, 19, 21, 86, 87, 90
causal sequence, *see* causal link
choice sequence, 29, 30, 32–35, 37, 41, 42, 78, 79, 81–83, 101, 102, 108, 115, 119, 120, 122, 125, 134, 135, 138, 145, 196–198, 201, 202
    break in the generation, 30, 33
    freedom of generation, 29, 31, 112, 121, 123, 125, 134, 145, 196
    randomness, 32, 82
    law-like (pre-determinate) sequence, 30, 81, 122, 145
    lawless sequence, 30, 122, 199
    restrictions on the generation, 29, 30, 82, 121, 198, 199
    second order restrictions, 31
classical provability, 181, 192
classical semantics, *see* two-valued semantics
color-word, 95, 178
communicability, *see* intersubjectivity
communication, *see* intersubjectivity
complete totality, 77, 78, 115, 120
compositionality, 172, 174, 185
comprehension axiom, 34, 40, 66
computable, *see* decidable, relation
consciousness
    Brouwer's concept, 18, 19
conservativeness, 52, 53
consistency, 52, 53
contextuality principle, 172–174
continuity principle, 79, 80, 200
    for disjunction, 80, 201
continuum, 28, 32, 33, 122, 203
    continuum vs. discreteness, 28

controllable sentence, 66, 72
counterexample
   σ-example, 73, 116, 118, 134
   Heyting's counterexamples, 113, 115–118
   strong, 33, 79, 81, 83, 197–202
   weak, 69–72, 77, 78, 117, 118

Davidson, 3
de Haen, 49
de Morgan's laws, 59
decidable
   relation
      provability as decidable relation, 13–16, 181
      sentence (in a formal theory), *see* undecidable, sentence (in a formal theory)
      statement, *see* yet-undecided statement or undecidable, truly undecidable statement or undecidable, sentence
Dedekind, 78
denotation, *see* reference
designated value, 153
determinism thesis, 74, 75
diagonal proof, 40
Dąmbska, 28

elimination rule, 190
embedding, *see* species, mathematical property
emotional content, *see* relations in a mathematical system
Erzeugungsprinzip, *see* ordinal number, Cantor's principles
Euclid, 54, 57, 58, 204
Euler's constant *C*, 116, 126
evidence, 66, 67, 71–74, 100, 119
   preserving evidence, 72–74
   stages of, 110–112
evident statement, 72
expectation, *see* intention
extensional relation, 79, 198, 200

Fermat's Last Theorem, 33, 46
fidelity, 159
Fine, 159

fitting-in, *see* species, mathematical property
force of an utterance, 6, 167, 170, 171
free choice sequence, *see* choice sequence
Frege, 3, 6, 9, 55, 59, 66, 85, 86, 107, 126, 152, 161, 165, 170, 172–174
frustration, *see* intention, frustration of intention
fully analyzed proof, 191
fundamental sequence, 36
future contingents, 75

generating operation, *see* species, addition of species
Goldbach's conjecture, 78, 185
Griss, 109–111, 145
Griss's schism, 109–112, 145
Grzegorczyk, 14–16
Gödel, 47, 120, 191
Gödel incompleteness theorem, 14, 47, 191, 193

Hilbert's Program, 13, 46, 48, 51–54, 83
   Brouwer's opposition to, 52–54
Hintikka, 152
holism, 173, 185, 186, 189, 192
Hume, 85
Humpty-Dumpty, 127, 135
Husserl, 85, 86, 127–132, 135, 144, 195
Hüttemann, 39

ideal meaning, 131
ideal sentence, 52
ideal truth, 66–68, 72
identity of content, 130
   as identity of species, 130
incommunicability, *see* intersubjectivity
indeterminacy, 74–79
indices, *see* spread, nodes, indices
infinity, 13, 64, 74–79, 83, 102, 115, 119, 120, 122, 145, 147, 148, 196, 197
   actual infinity, 51, 76, 102, 123

potential infinity, 76, 102, 122, 134, 135, 145
ingredients of meaning, 165, 173, 181–185
intensional function, 197, 199
intention, 127
　frustration of intention, 131, 132
　fulfillment of intention, 126, 127, 129, 131
　meaning intention, 129
　non-fulfillment of intention, 131, 135, 145
interpretation
　by replacing schematic letters, 151, 152
　internal interpretation, 156, 157
　interpretation vs. meaning, 162, 164
　programmatic interpretation, 155, 156, 158, 160, 161
intersubjectivity, 1–5, 7–9, 193
　formalist's guarantee of intersubjectivity, 13–16
　in Brouwer's philosophy, 100
　in Heyting's doctrine, 144
　mathematicians on intersubjectivity, 12–16
　mentalist condition, 4, 5, 10, 100, 194
　　in Brouwer's doctrine, 90
　　in Heyting's conception, 140, 142–144
　Wittgensteinian condition, 9–12, 96, 100, 142, 144, 194
　　in Brouwer's doctrine, 95
　　in Heyting's conception, 140, 142, 144
intimation, 128, 129
intrasubjective reality, 203
intuition, 15, 17
　as non-inferential knowing, 41
　as self-evidence, 112
　intuition of time, 41
　intuition of two-ity, 19, 27–29, 36, 40, 51, 64, 67, 87, 88, 100, 103, 144
　　and mathematics, 27, 29–36
　　as a defining property of mind, 43, 88, 139
　　objections to, 41–44
　intuition of two-oneness, 27
　Kant's concept of intuition, 28, 88
intuitive fullness, 128

Jevons, 60

Kant, 28, 41, 43, 44, 88, 89, 103, 104, 106, 129
Klaus, 5
Kolmogorov, 135, 136
Korteweg, 17, 62, 63, 203
Kripke, 153, 154, 156
Kutsch, 204

language
　accompanying mathematical constructions, 40, 52, 56, 61, 68, 69, 98
　as a means of supporting the memory, 51, 96, 140
　as an imperfect means of communication, 15, 51
　linguistic means of the introduction of mathematical objects, 40, 48, 64, 66
　mastery of language
　　as a practical ability, 170
　　as theoretical knowledge, 170
　representation of mathematical reasoning, 50, 55, 56
　social account of language-learning, 91, 141
　social character of language, 167, 182
　with vagueness, 161
law-like sequence, see choice sequence
lawless sequence, see choice sequence
Leibniz, 55, 62
Leśniewski, 13, 14
linguistic causal attention, 49, 144
Locke, 50, 85, 140
logic
　Brouwer's concept of, 40, 59–62, 64, 65, 69
　apriority of laws of logic, 60, 65
　origin of logic, 61

Heyting's views on logic, 137–139
laws of logic
    as laws of thought, 85
    as preserving evidence, 72–74
    counterexamples, see counterexample
    logic vs. mathematics, see mathematics vs. logic
logical constants
    in terms of assertability, 136
    in terms of intentions, 127
    in terms of provability, 136, 137
    Kolmogorov's account, 135, 136
logical objects, 40, 64
logicism, 60, 107
logistic, 62

Łukasiewicz, 74, 75, 85, 86

manifestability, see truth conditions, manifestability
Mannoury, 49, 99
Martin-Löf, 3, 204
mathematical abstraction, 27, 100, 106
mathematical attention, 49, 61
mathematical entities, 29, 34, 35, 40, 104, 107
mathematical system, see relations in a mathematical system
mathematics vs. logic, 59, 62–65
McGinn, 205
meaning that transcends use, 181–185
meaning theory, 165, 169–171, 173, 174, 176, 180, 184, 186–188, 192–195
meaning vs. understanding, 167
meanings of logical constants, see logical constants
metalanguage, 154, 156
methodological principle, 186
mind
    Brouwer's concept of mind, 19, 27, 41, 89, 90
    other minds, 22–27, 89
model theory, 120, 155, 175
Murawski, 204

natural deduction, 70, 190

natural science
    Brouwer's views, 22
negationless mathematics, see Griss's schism
nested intervals, 116
neutrality argument, 108, 112, 113
non-statable knowledge, 177, 205
normalized proof, 190

object
    Brouwer's concept of, 21, 86, 87, 90
objective expression (Husserl), 130
ontological argument, see neutrality argument
order relations (intuitionistic), 36, 37, 116–118
ordinal number, 36–39, 64
    Cantor's principles, 39, 40
Ornstein, 49
other minds, see mind

Peano, 54, 60, 62, 120
Peano arithmetic, 53, 54, 120
Peirce, 60
Platonist assumption, 104, 107, 112, 119, 122, 196
possibility
    Brouwer's notion of, 44, 45, 48, 69, 70
Posy, 203
Prawitz, 205
pre-determinate, see choice sequence
precisification, 159, 160
principle of access, 133
principle of data, 198, 200
principle of double negation elimination, 57, 59, 70, 83, 154
    generalized, 79
principle of identity, 59
principle of non-contradiction, 59
principle of the excluded middle, 33, 46, 47, 59, 70–72, 81, 105, 115–118, 120, 127, 154, 155, 159
    generalized, 79, 83, 201, 202
    temporal, 74
private language argument, 50, 92
private naming, 10, 93, 94, 102, 140, 142

INDEX 217

private ostensive definition, 10, 13, 91, 96, 100, 140
problem-solving, *see* logical constants, Kolmogorov's account
proof theoretical argument, 205
property, *see* species, mathematical property
provability
    as a central notion of meaning theory, 189
    as a decidable relation, 187, 188
psychological argument, 107, 109–112
psychologism, 2, 84, 86, 89, 90, 102, 106
    epistemological thesis, 84, 85
    genealogical thesis, 85, 89
    ontological thesis, 84

random numbers, 32, 82
real number, 28, 30, 33, 79, 112, 116–118, 122
    Dedekind's definition, 116, 117
    intuitionistic definition, 116, 117
    order relations, 117, 118
    real number generator, 117, 118
        coincidence, 117
real sentence, 52
reference, 5, 7–10, 20, 21, 53, 66, 91, 92, 95–97, 100, 142, 166, 195
    referring to mental objects, 90–95, 97, 143, 144
regularities in language, *see* logic, Brouwer's concept of
relations in a mathematical system, 86, 89
    representations (subjective), 86, 89
restrictions, *see* choice sequence
routine practice, 8
rule of double negation elimination, 70, 71–73, 105, 118, 135, 188
Russell, 60, 62, 107, 119

schematic letters, 151
Schopenhauer, 17, 18
Schröder, 60
second number class, 36, 38–40
second order restrictions, *see* choice sequence

semantic theory, 144, 149–151, 153–156, 158–161, 163, 164, 171, 173–175, 187, 192, 196
    genuine semantic theory, 153
    sensitive semantic theory, 155
    three tasks of semantic theory, 155
semantic value, 152, 160, 173, 174
semantical argument, 108, 126–136
sense, 165, 170, 171, 173
    determination of semantic value, 173, 174, 176
sense-fulfilling acts, 129
sense-giving acts, 129
sentence vs. statement (Heyting's account), 126
Significs, 49, 68, 96
Skolem-Löwenheim theorem, 14
Socrates, 204
solipsism, 2, 22, 26, 27, 84, 87, 89, 102
Sophists, 12
species, 28, 29, 36
    addition of species, 36, 37
    completely ordered species, 36
    mathematical property, 35
    species of higher order, 35
    well ordered species, 37
spread, 28, 34, 35, 108, 122, 191
    nodes, 34, 163, 164
        descendant of a node, 34
        indices, 34
    spread-law, 34
    sterilization, 33
stability, 159
statable knowledge, 177, 205
statement value, 152, 153, 155, 158
    and truth, 153, 155
straightforward stipulation, 155, 156, 158
strong counterexample, *see* counterexample
subjective expression (Husserl), 130
subjectivism, 2, 84, 87–89
Sundholm, 177
super-subject, 25, 68
Swart de, 162, 163
syllogism, 59, 60, 204

Tait, 203

Tarski, 3
temporal attention, 19, 22, 27, 144
theory of meaning, 165
Tieszen, 41
transcendental subject, 43, 44, 88, 89, 106
transmission of will, *see* will-transmission
transubjective (or transient) reality, 204
Troelstra, 203
truth, 2–4, 18, 21, 25, 26, 44, 58, 65–69, 71, 72, 74, 75, 77, 78, 83, 99–102, 105, 107, 108, 136, 142, 144, 145, 151–153, 155–161, 165, 171, 172, 175, 176, 179
  and statement value, 152
  and vagueness, 160
  at a node, 156, 157, 162
  at a time, 75
  Brouwer's concept of, 67–69, 99
  constructive concept of, 76, 78
  verification-transcendent, 74, 76, 136
truth conditions, 176–179, 181, 183–185, 187, 195
  manifestability, 181, 186, 187, 193
truth vs. meaning, 3
truth-value, 158, 173, 178
Turnau, 203
twin numbers, 44, 114, 115, 120, 121
two-valued semantics, 171, 175, 176, 178, 181, 193, 194

undecidable
  truly undecidable statement, 33, 46–48, 71, 75, 77, 184
  undecidable sentence (in a formal theory), 15, 16, 47, 48, 120, 121
understanding, *see* meaning vs. understanding

vagueness, *see* language, with vagueness, 161
van Dalen, 136, 203, 204
van Eeden, 49
van Stigt, 67, 86, 137, 203, 204

weak counterexample, *see* counterexample
Weyl, 28
Wiles, 46
will-transmission, 2, 49, 50, 65, 97–100
Wittgenstein, v, 1, 9, 10, 12, 42, 50, 92–94, 96, 100, 140–142, 144, 161, 171, 181, 194

yet-undecided statement, 46, 47, 71, 73, 75, 76, 114, 116, 119, 124, 165, 177, 179, 180, 183–185, 193

Zermello, 34, 48

SYNTHESE LIBRARY

243. G. Debrock and M. Hulswit (eds.), *Living Doubt*. Essays concerning the epistemology of Charles Sanders Peirce. 1994　ISBN 0-7923-2898-1
244. J. Srzednicki, *To Know or Not to Know*. Beyond Realism and Anti-Realism. 1994
　ISBN 0-7923-2909-0
245. R. Egidi (ed.), *Wittgenstein: Mind and Language*. 1995　ISBN 0-7923-3171-0
246. A. Hyslop, *Other Minds*. 1995　ISBN 0-7923-3245-8
247. L. Pólos and M. Masuch (eds.), *Applied Logic: How, What and Why*. Logical Approaches to Natural Language. 1995　ISBN 0-7923-3432-9
248. M. Krynicki, M. Mostowski and L.M. Szczerba (eds.), *Quantifiers: Logics, Models and Computation*. Volume One: Surveys. 1995　ISBN 0-7923-3448-5
249. M. Krynicki, M. Mostowski and L.M. Szczerba (eds.), *Quantifiers: Logics, Models and Computation*. Volume Two: Contributions. 1995　ISBN 0-7923-3449-3
　Set ISBN (Vols 248 + 249) 0-7923-3450-7
250. R.A. Watson, *Representational Ideas from Plato to Patricia Churchland*. 1995
　ISBN 0-7923-3453-1
251. J. Hintikka (ed.), *From Dedekind to Gödel*. Essays on the Development of the Foundations of Mathematics. 1995　ISBN 0-7923-3484-1
252. A. Wiśniewski, *The Posing of Questions*. Logical Foundations of Erotetic Inferences. 1995
　ISBN 0-7923-3637-2
253. J. Peregrin, *Doing Worlds with Words*. Formal Semantics without Formal Metaphysics. 1995
　ISBN 0-7923-3742-5
254. I.A. Kieseppä, *Truthlikeness for Multidimensional, Quantitative Cognitive Problems*. 1996
　ISBN 0-7923-4005-1
255. P. Hugly and C. Sayward: *Intensionality and Truth*. An Essay on the Philosophy of A.N. Prior. 1996　ISBN 0-7923-4119-8
256. L. Hankinson Nelson and J. Nelson (eds.): *Feminism, Science, and the Philosophy of Science*. 1997　ISBN 0-7923-4162-7
257. P.I. Bystrov and V.N. Sadovsky (eds.): *Philosophical Logic and Logical Philosophy*. Essays in Honour of Vladimir A. Smirnov. 1996　ISBN 0-7923-4270-4
258. Å.E. Andersson and N-E. Sahlin (eds.): *The Complexity of Creativity*. 1996
　ISBN 0-7923-4346-8
259. M.L. Dalla Chiara, K. Doets, D. Mundici and J. van Benthem (eds.): *Logic and Scientific Methods*. Volume One of the Tenth International Congress of Logic, Methodology and Philosophy of Science, Florence, August 1995. 1997　ISBN 0-7923-4383-2
260. M.L. Dalla Chiara, K. Doets, D. Mundici and J. van Benthem (eds.): *Structures and Norms in Science*. Volume Two of the Tenth International Congress of Logic, Methodology and Philosophy of Science, Florence, August 1995. 1997　ISBN 0-7923-4384-0
　Set ISBN (Vols 259 + 260) 0-7923-4385-9
261. A. Chakrabarti: *Denying Existence*. The Logic, Epistemology and Pragmatics of Negative Existentials and Fictional Discourse. 1997　ISBN 0-7923-4388-3
262. A. Biletzki: *Talking Wolves*. Thomas Hobbes on the Language of Politics and the Politics of Language. 1997　ISBN 0-7923-4425-1
263. D. Nute (ed.): *Defeasible Deontic Logic*. 1997　ISBN 0-7923-4630-0
264. U. Meixner: *Axiomatic Formal Ontology*. 1997　ISBN 0-7923-4747-X
265. I. Brinck: *The Indexical 'I'*. The First Person in Thought and Language. 1997
　ISBN 0-7923-4741-2
266. G. Hölmström-Hintikka and R. Tuomela (eds.): *Contemporary Action Theory*. Volume 1: Individual Action. 1997　ISBN 0-7923-4753-6; Set: 0-7923-4754-4

SYNTHESE LIBRARY

267. G. Hölmström-Hintikka and R. Tuomela (eds.): *Contemporary Action Theory.* Volume 2: Social Action. 1997 ISBN 0-7923-4752-8; Set: 0-7923-4754-4
268. B.-C. Park: *Phenomenological Aspects of Wittgenstein's Philosophy.* 1998
 ISBN 0-7923-4813-3
269. J. Pasśniczek: *The Logic of Intentional Objects.* A Meinongian Version of Classical Logic. 1998 ISBN 0-7923-4880-X
270. P.W. Humphreys and J.H. Fetzer (eds.): *The New Theory of Reference.* Kripke, Marcus, and Its Origins. 1998 ISBN 0-7923-4898-2
271. K. Szaniawski, A. Chmielewski and J. Woleński (eds.): *On Science, Inference, Information and Decision Making.* Selected Essays in the Philosophy of Science. 1998
 ISBN 0-7923-4922-9
272. G.H. von Wright: *In the Shadow of Descartes.* Essays in the Philosophy of Mind. 1998
 ISBN 0-7923-4992-X
273. K. Kijania-Placek and J. Woleński (eds.): *The Lvov–Warsaw School and Contemporary Philosophy.* 1998 ISBN 0-7923-5105-3
274. D. Dedrick: *Naming the Rainbow.* Colour Language, Colour Science, and Culture. 1998
 ISBN 0-7923-5239-4
275. L. Albertazzi (ed.): *Shapes of Forms.* From Gestalt Psychology and Phenomenology to Ontology and Mathematics. 1999 ISBN 0-7923-5246-7
276. P. Fletcher: *Truth, Proof and Infinity.* A Theory of Constructions and Constructive Reasoning. 1998 ISBN 0-7923-5262-9
277. M. Fitting and R.L. Mendelsohn (eds.): *First-Order Modal Logic.* 1998
 ISBN 0-7923-5334-X
278. J.N. Mohanty: *Logic, Truth and Modalities.* From a Phenomenological Perspective. 1999
 ISBN 0-7923-5550-4
279. T. Placek: *Mathematical Intuitionism and Intersubjectivity.* A Critical Exposition of Arguments for Intuitionism. 1999 ISBN 0-7923-5630-6
280. A. Cantini, E. Casari and P. Minari (eds.): *Logic and Foundations of Mathematics.* 1999
 ISBN 0-7923-5659-4
281. M.L. Dalla Chiara, R. Giuntini and F. Laudisa (eds.): *Language, Quantum, Music.* 1999
 ISBN 0-7923-5727-2
282. P. Fletcher: *Truth, Proof and Infinity.* A Theory of Constructions and Constructive Reasoning. 1998 ISBN 0-7923-5262-9
283. M. Fitting and R.L. Mendelsohn: *First-Order Modal Logic.* 1998
 ISBN 0-7923-5334-X (HB); ISBN 0-7923-5335-8 (PB)
284. J.N. Mohanty: *Logic, Truth and the Modalities from a Phenomenological Perspective.* 1998
 ISBN 0-7923-5550-4

Previous volumes are still available.

KLUWER ACADEMIC PUBLISHERS – DORDRECHT / BOSTON / LONDON